High-Speed Pulse Circuits

High-Speed Pulse Circuits

Arpad Barna
University of Hawaii

WILEY–INTERSCIENCE
A Division of John Wiley & Sons, Inc.
New York London Sydney Toronto

Library of Congress Catalog Card Number: 76-121904
ISBN 0-471-05033-4

Printed in the United States of America

10 9 8 7 6 5 4 3 2 1

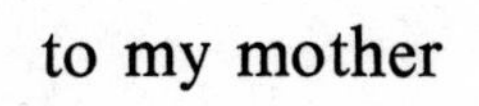

to my mother

Preface

Pulse circuits have found extensive and growing applications in many fields, such as industrial electronics, nuclear electronics, digital computers, digital systems; and there are several excellent texts available dealing with various aspects of the subject.

The purpose of this book is to provide an introduction to *high-speed* pulse circuits with particular emphasis on circuits utilizing semiconductors; it is aimed at the engineer and scientist engaged in analysis, design, evaluation, or application, and at the student studying circuit analysis and design. Although the presentation does not rely on prior studies in circuit theory, transmission line theory, or semiconductor devices, some background in these fields is helpful; a background in calculus is essential.

The overall goal has been to emphasize techniques used in high-speed circuits and, at the same time, provide material for use in a senior level course. Over 100 exercises are included, some of them with answers provided. These enlarge upon the material presented in the text and demonstrate applications.

Chapter 1 describes ideal components, and simple transients in transmission lines are discussed. In Chapter 2, the Laplace transform as a basic tool of linear transient analysis is introduced and applied to the analysis of transients in *R-L-C* circuits and in transmission lines with arbitrary resistive terminations. Risetimes of cascaded circuits are discussed in terms of the Elmore risetime in Chapter 3, and digital computer techniques are introduced in the analysis of an *L-C* ladder network. In Chapter 4, limitations of real resistors, capacitors, inductors, and transformers in high-speed pulse circuits are summarized, characteristics of practical transmission lines are described, and pulse degradation is discussed.

Nonlinear components are introduced in Chapter 5 dealing with junction diodes. A simple model incorporating conductivity modulation is described, and transients in junction diode circuits are analyzed by means of a digital computer. In Chapter 6, tunnel diode models are described and utilized in the computation of risetime and ringing in tunnel diode circuits.

The transistor and its high-speed parameters are introduced in Chapter 7, and the transient response of grounded-base, grounded-emitter, and grounded-collector configurations are derived. Switching properties of the emitter-coupled pair are analyzed in detail by the use of digital computation techniques in Chapter 8. Transient response and optimization of high-speed digital fanout circuits are discussed in Chapter 9. A high-speed amplifier suitable for integrated-circuit construction is described in Chapter 10.

The material presented here grew out of a seminar series at the Stanford Linear Accelerator Center of Stanford University. I am deeply indebted to many of my colleagues for their comments, suggestions, and encouragement, and also to the Stanford Linear Accelerator Center for providing the creative atmosphere that made this book possible.

Arpad Barna

Honolulu, Hawaii
April 1970

Contents

1

Ideal Components

Basic properties of ideal resistors, capacitors, inductors, voltage and current sources, switches, transformers, and transmission lines are summarized in this chapter. These components are linear and are characterized by a single quantity, such as resistance, capacitance, etc. In contrast, real components—discussed in detail in Chapter 4—can be considered as combinations of ideal components and are described by several quantities.*

THE IDEAL RESISTOR

The ideal resistor is a two-terminal device (Fig. 1.1).† The relationship between the voltage $V(t)$ across and the current $I(t)$ through its terminals is given by

$$V(t) = RI(t), \tag{1.1}$$

* More details may be found in the references.
† Here, and in what follows, voltage arrows point from the negative terminal to the positive one, and current arrows along the direction of the current.

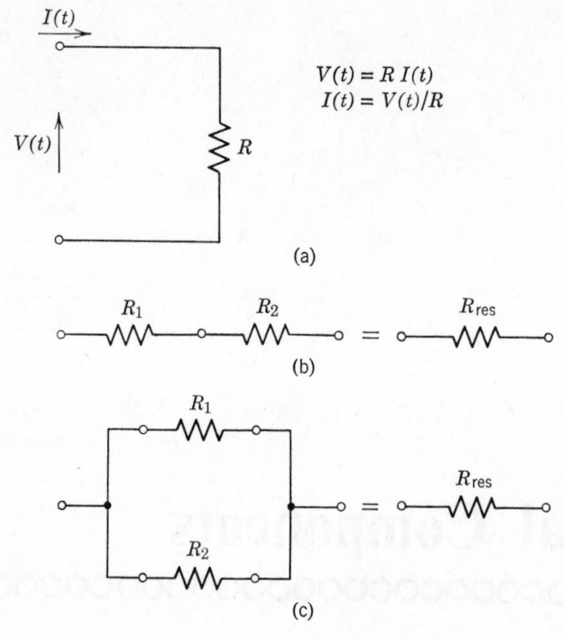

Figure 1.1 The ideal resistor. (a) Terminal relations. (b) Series connection: $R_{\text{res}} = R_1 + R_2$. (c) Parallel connection: $R_{\text{res}} = R_1R_2/(R_1 + R_2)$.

where R is the resistance in ohms: 1 ohm = 1 Ω = 1 volt/ampere = 1 V/A. Also used are the units mΩ = 10^{-3} Ω, kΩ = 10^3 Ω, MΩ = 10^6 Ω. Thus, a 1-Ω resistor has 1 volt across its terminals when a current of 1 ampere flows through it.

The ideal resistor cannot store energy. It is a purely dissipative element, converting into heat a power of

$$P(t) = V(t)I(t) = \frac{V^2(t)}{R} = I^2(t)R. \tag{1.2}$$

When two resistors R_1 and R_2 are connected in series, the resulting resistance is $R_{\text{res}} = R_1 + R_2$. The resistance R_{res} resulting by connecting R_1 and R_2 in parallel is $R_{\text{res}} = R_1R_2/(R_1 + R_2)$, also $1/R_{\text{res}} = 1/R_1 + 1/R_2$. If m resistors are connected in series, the resulting resistance is $R_{\text{res}} = R_1 + R_2 + \cdots + R_m$. If m resistors are connected in parallel, then $1/R_{\text{res}} = 1/R_1 + 1/R_2 + \cdots + 1/R_m$.

THE IDEAL CAPACITOR

The ideal capacitor is a two-terminal device (Fig. 1.2). The relationship between the voltage $V(t)$ across and the current $I(t)$ through its terminals is given by

$$I(t) = C \frac{dV(t)}{dt}, \tag{1.3}$$

or by

$$V(t) = \int_{-\infty}^{t} \frac{1}{C} I(t')\, dt', \tag{1.4}$$

where C is the capacitance measured in farads: 1 farad = 1 F = 1 second/ohm = 1 s/Ω = 1 As/V = 1 coulomb/V. Also used are the units pF = 10^{-12} F, nF = 10^{-9} F, μF = 10^{-6} F. The voltage across a 1-farad capacitor changes 1 volt after a current of 1 ampere

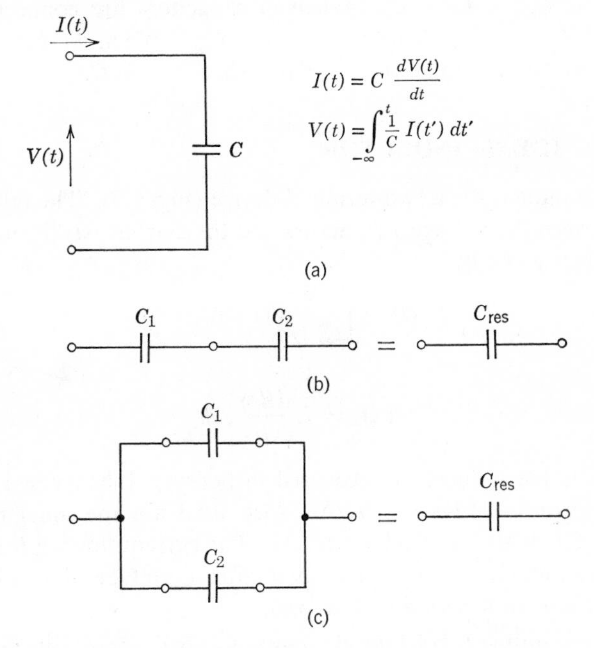

Figure 1.2 The ideal capacitor. (a) Terminal relations. (b) Series connection: $C_{res} = C_1C_2/(C_1 + C_2)$. (c) Parallel connection: $C_{res} = C_1 + C_2$.

has been applied for a period of 1 second. If the capacitance is constant, Equation (1.4) can also be written as

$$V(t) = \frac{1}{C}\int_{-\infty}^{t} I(t')\, dt'. \tag{1.5}$$

The ideal capacitor is a non-dissipative element. A constant capacitor C charged to a voltage V stores an energy of E contained in an electric field:

$$E = \tfrac{1}{2}CV^2, \tag{1.6}$$

and a charge of

$$Q = CV. \tag{1.7}$$

When two capacitors C_1 and C_2 are connected in series, the resulting capacitance is $C_{res} = C_1C_2/(C_1 + C_2)$, also $1/C_{res} = 1/C_1 + 1/C_2$. The capacitance resulting by connecting C_1 and C_2 in parallel is $C_{res} = C_1 + C_2$. When m capacitors are connected in series, $1/C_{res} = 1/C_1 + 1/C_2 + \cdots + 1/C_m$. When m capacitors are connected in parallel, $C_{res} = C_1 + C_2 + \cdots + C_m$.

THE IDEAL INDUCTOR

The ideal inductor is a two-terminal device (Fig. 1.3). The relationship between the voltage $V(t)$ across and the current $I(t)$ through its terminals is given by

$$I(t) = \int_{-\infty}^{t} \frac{1}{L} V(t')\, dt', \tag{1.8}$$

or by

$$V(t) = L\frac{dI(t)}{dt}, \tag{1.9}$$

where L is the inductance measured in henrys: 1 henry = 1 H = 1 ohm second = 1 Ωs = 1 Vs/A. Also used are the units nH = 10^{-9} H, μH = 10^{-6} H, mH = 10^{-3} H. The current flowing through a 1-henry inductor changes 1 ampere after a voltage of 1 volt has been applied for a period of 1 second.

The ideal inductor is a non-dissipative element. When the current in it is I, it stores an energy of E contained in a magnetic field:

$$E = \tfrac{1}{2}LI^2, \tag{1.10}$$

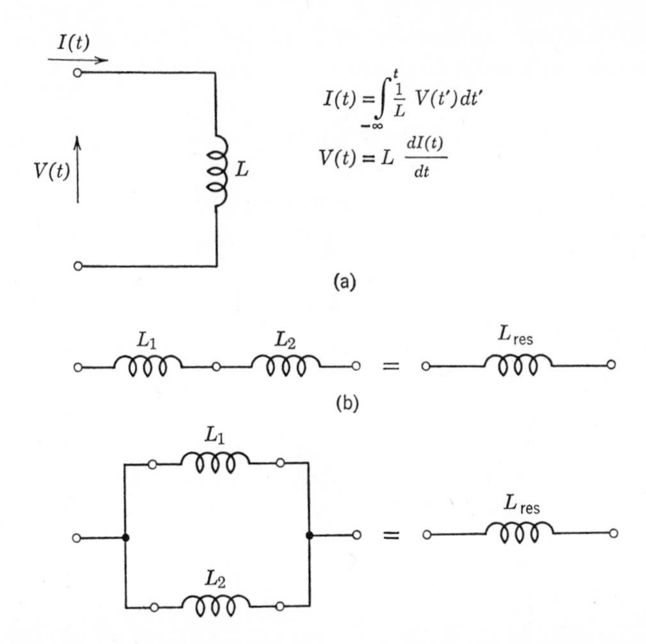

Figure 1.3 The ideal inductor. (a) Terminal relations. (b) Series connection: $L_{res} = L_1 + L_2$. (c) Parallel connection: $L_{res} = L_1L_2/(L_1 + L_2)$.

and a magnetic flux of

$$\Phi = LI. \tag{1.11}$$

When two inductors L_1 and L_2 are connected in series, the resulting inductance is $L_{res} = L_1 + L_2$. The parallel connection of L_1 and L_2 results in $L_{res} = L_1L_2/(L_1 + L_2)$, also $1/L_{res} = 1/L_1 + 1/L_2$. In general, when m inductors are connected in series, $L_{res} = L_1 + L_2 + \cdots + L_m$. When m inductors are connected in parallel, $1/L_{res} = 1/L_1 + 1/L_2 + \cdots + 1/L_m$.

THE IDEAL VOLTAGE SOURCE

The ideal voltage source (Fig. 1.4) can be a source of power. It supplies a voltage $V_g(t)$ independent of current as long as the load is

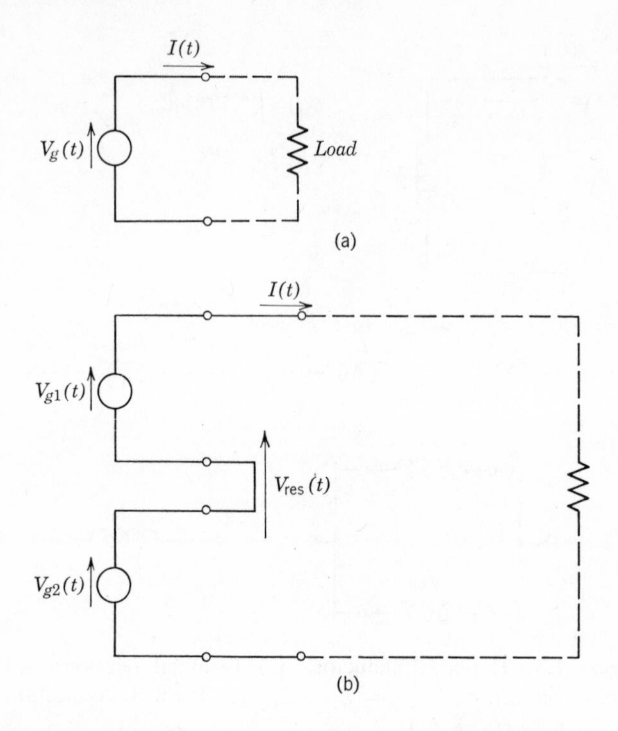

Figure 1.4 The ideal voltage source. (a) Symbol. (b) Series connection: $V_{\mathrm{res}}(t) = V_{g1}(t) + V_{g2}(t)$.

restricted such that the terminal current $I(t)$ is finite for a finite source voltage $V_g(t)$. It delivers a power of

$$P(t) = V_g(t)I(t). \tag{1.12}$$

The unit of the voltage is the volt (V). Also used are the units nV = 10^{-9} V, μV = 10^{-6} V, mV = 10^{-3} V, kV = 10^3 V, MV = 10^6 V, GV = 10^9 V.

When two voltage sources $V_{g1}(t)$ and $V_{g2}(t)$ are connected in series, the resulting voltage source is $V_{\mathrm{res}}(t) = V_{g1}(t) + V_{g2}(t)$. When m voltage sources are connected in series, the resulting voltage source is $V_{\mathrm{res}}(t) = V_{g1}(t) + V_{g2}(t) + \cdots + V_{gm}(t)$. Parallel connection of voltage sources is not allowed since, unless they are identical, this would result in an infinite current.

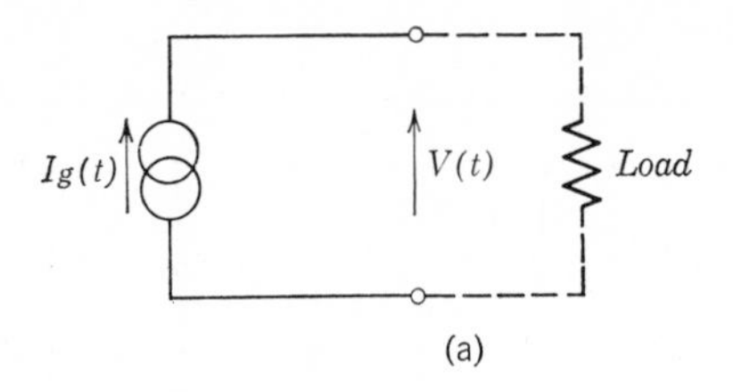

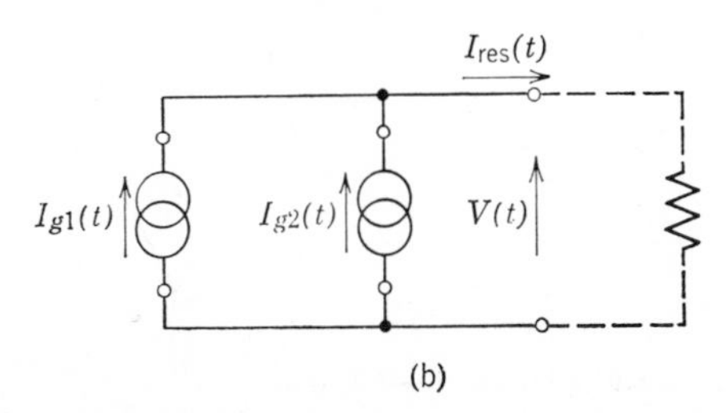

Figure 1.5 The ideal current source. (a) Symbol. (b) Parallel connection: $I_{\text{res}}(t) = I_{g1}(t) + I_{g2}(t)$.

THE IDEAL CURRENT SOURCE

The ideal current source (Fig. 1.5) can be a source of power. It supplies a current $I_g(t)$ independent of voltage as long as the load is restricted such that the terminal voltage $V(t)$ is finite for a finite source current $I_g(t)$. It delivers a power of

$$P(t) = V(t)I_g(t). \tag{1.13}$$

The unit of the current is the ampere (A). Also used are the units $\text{pA} = 10^{-12}\text{ A}$, $\text{nA} = 10^{-9}\text{ A}$, $\mu\text{A} = 10^{-6}\text{ A}$, $\text{mA} = 10^{-3}\text{ A}$, $\text{kA} = 10^{3}\text{ A}$.

When two current sources $I_{g1}(t)$ and $I_{g2}(t)$ are connected in parallel, the resulting current source is $I_{\text{res}}(t) = I_{g1}(t) + I_{g2}(t)$. When m current sources are connected in parallel, the resulting current source is $I_{\text{res}}(t) = I_{g1}(t) + I_{g2}(t) + I_{g3}(t) + \cdots + I_{gm}(t)$. Series connection of current sources is not allowed since, unless they are identical, this would result in an infinite voltage.

THE IDEAL SWITCH

The ideal switch (Fig. 1.6) is a two-terminal network which can have two states: *Closed* and *Open*. In its *Closed* state it is equivalent to

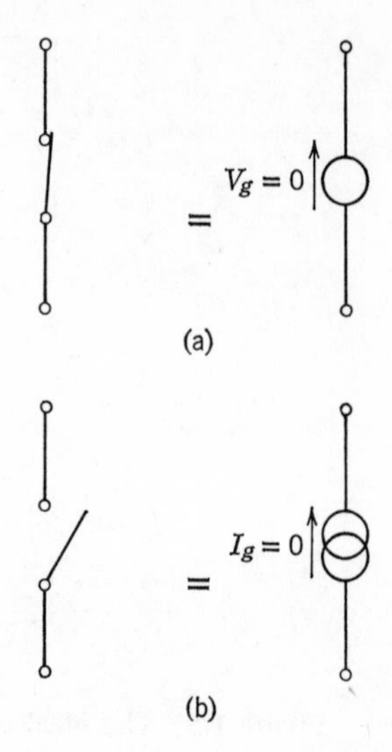

Figure 1.6 The ideal switch. (a) *Closed.* (b) *Open.*

an ideal voltage source with $V_g = 0$; in its *Open* state it is equivalent to an ideal current source with $I_g = 0$.

THE IDEAL TRANSFORMER

The ideal transformer [Fig. 1.7(a)] is a four-terminal network characterized by the "turns-ratio" N, a dimensionless number. The relationships between the voltages and the currents are

$$V_2(t) = NV_1(t) \tag{1.14}$$

and

$$I_2(t) = \frac{I_1(t)}{N}. \tag{1.15}$$

From these it follows that

$$V_2(t)I_2(t) = V_1(t)I_1(t). \tag{1.16}$$

Thus, the ideal transformer is non-dissipative and it is not a source of power; within the restriction of Equation (1.16), it can transform voltages and currents up and down.

Parallel-series connection of transformers is shown in Fig. 1.7(b). Here

$$V_2(t) = V_1(t)(N_A + N_B) \tag{1.17}$$

and

$$I_2(t) = \frac{I_1(t)}{N_A + N_B}\,. \tag{1.18}$$

THE IDEAL TRANSMISSION LINE

The ideal transmission line (Fig. 1.8) can transmit signals in either direction without distortion. A signal introduced at the left pair of its terminals travels to the right and reaches the right pair of

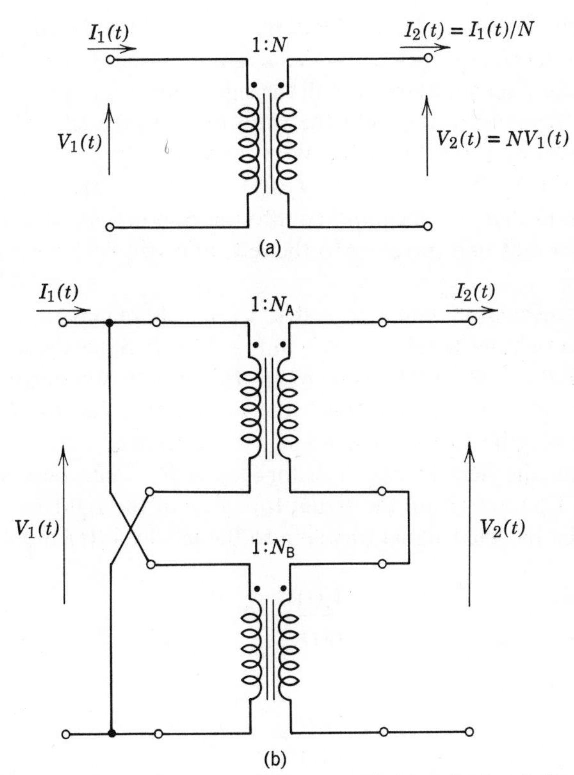

Figure 1.7 The ideal transformer. (a) Terminal relations. (b) Parallel-series connection: $V_2(t) = V_1(t)(N_A + N_B$, $I_2(t) = I_1(t)/(N_A + N_B)$.

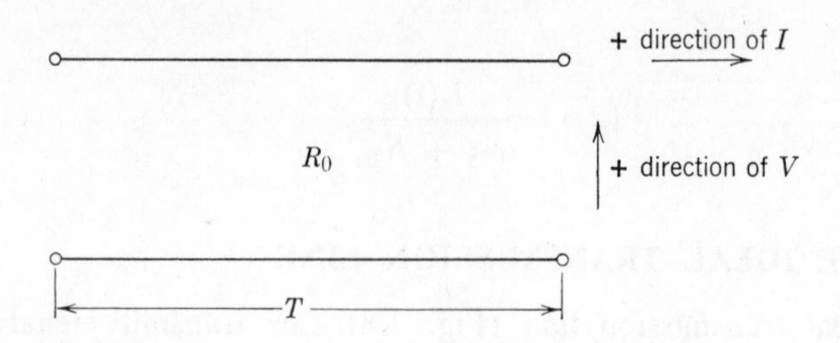

Figure 1.8 Ideal transmission line with delay T and characteristic impedance R_0.

terminals T seconds later. Similarly, a signal introduced at the right pair of terminals travels to the left and reaches the left pair of terminals T seconds later. With the sign conventions as shown, for a signal traveling to the right the ratio between the traveling voltage and traveling current is $+R_0$. With the same sign conventions, for a signal traveling to the left $V(t)/I(t) = -R_0$. Signals on a transmission line can be decomposed into the sum of a signal traveling to the right and one traveling to the left, and will be done so in what follows.

A transmission line terminated by its characteristic impedance R_0 at its right terminals is shown in Fig. 1.9(a). Since the terminating resistor $R_T = R_0$, $V(t)/I(t) = R_T = R_0$. Hence, the signal traveling to the right cannot distinguish the terminating resistor R_0 from a further length of cable, and no signal is reflected.

When the terminating resistor $R_T \neq R_0$, reflection will take place. Characterizing the signal traveling to the right by $V_i(t)$ and $I_i(t)$, the reflected signal traveling to the left by $V_r(t)$ and $I_r(t)$,

$$\frac{V_i(t)}{I_i(t)} = R_0 \tag{1.19}$$

and

$$\frac{V_r(t)}{I_r(t)} = -R_0. \tag{1.20}$$

Figure 1.9(b) shows a transmission line terminated by an open circuit, $R_T = \infty$, with the line initially uncharged, i.e., with

$V(t < 0) = 0$. Since the current into the terminating resistor of $R_T = \infty$ is zero,

$$I_i(t) + I_r(t) = 0. \tag{1.21}$$

Combining Equations (1.19), (1.20), and (1.21) results in

$$\frac{V_r(t)}{V_i(t)} = +1, \tag{1.22}$$

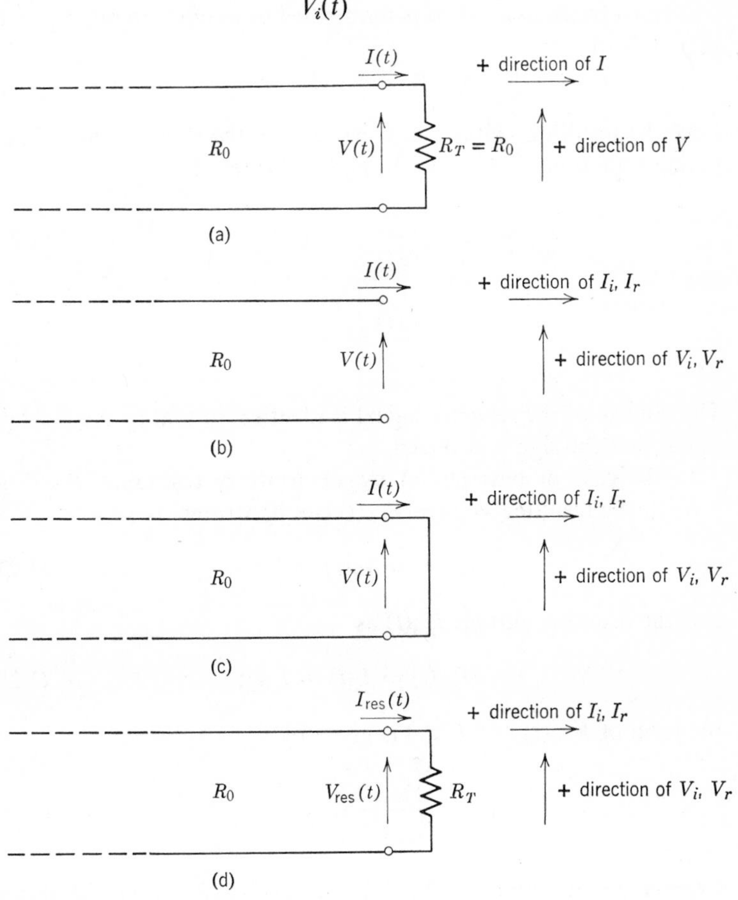

Figure 1.9 Transmission line terminated by: (a) Its characteristic impedance. (b) An open circuit. (c) A short circuit. (d) An arbitrary resistance R_T.

and

$$\frac{I_r(t)}{I_i(t)} = -1. \tag{1.23}$$

Thus, the voltage of the reflected signal is identical to the incident signal voltage in magnitude and in sign.

When a transmission line is terminated by a short-circuit, $R_T = 0$, [Fig. 1.9(c)]:

$$V_i(t) + V_r(t) = 0, \tag{1.24}$$

since the resulting voltage must be zero on the short-circuit. Combining Equations (1.19), (1.20), and (1.24) results in

$$\frac{V_r(t)}{V_i(t)} = -1 \tag{1.25}$$

and

$$\frac{I_r(t)}{I_i(t)} = +1. \tag{1.26}$$

The voltage of the reflected signal is identical to that of the incident signal, but the sign is reversed.

In the case of termination by an arbitrary resistance R_T [Fig. 1.9(d)], the resulting voltage $V_{\text{res}}(t)$ can be written as

$$V_{\text{res}}(t) = V_i(t) + V_r(t), \tag{1.27}$$

and the resulting current $I_{\text{res}}(t)$ as

$$I_{\text{res}}(t) = I_i(t) + I_r(t); \tag{1.28}$$

the ratio of $V_{\text{res}}(t)$ and $I_{\text{res}}(t)$ is given by

$$\frac{V_{\text{res}}(t)}{I_{\text{res}}(t)} = R_T. \tag{1.29}$$

Combining Equations (1.19), (1.20), (1.27), (1.28), and (1.29) results in

$$\frac{V_r(t)}{V_i(t)} = \frac{R_T - R_0}{R_T + R_0} \tag{1.30}$$

and

$$\frac{I_r(t)}{I_i(t)} = -\frac{R_T - R_0}{R_T + R_0}, \tag{1.31}$$

where the quantity $(R_T - R_0)/(R_T + R_0)$ is called the reflection coefficient, often denoted by ρ.

EXERCISES

1. Show that in the circuit of Fig. 1.10

$$V(t) = V_g(t)\frac{R_2}{R_1 + R_2}.$$

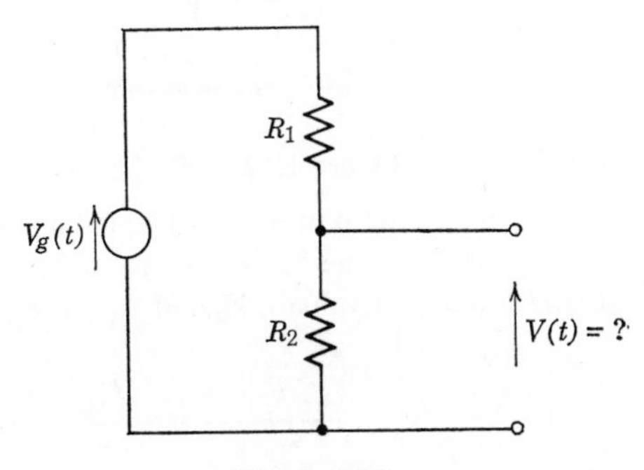

Figure 1.10

2. Show that in the circuit of Fig. 1.11

$$I_2(t) = I_g(t)\frac{R_1}{R_1 + R_2}.$$

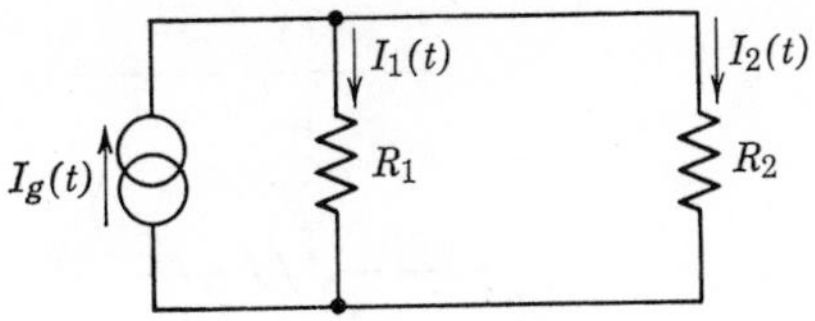

Figure 1.11

3. Show that, as observed by $V(t)$ and $I(t)$ on the terminals, the circuit of Fig. 1.12(a) ("Thevenin equivalent circuit") is equivalent to that of Fig. 1.12(b) ("Norton equivalent circuit") if, and only if, $r_g = R_g$ and $I_g(t) = V_g(t)/R_g$.

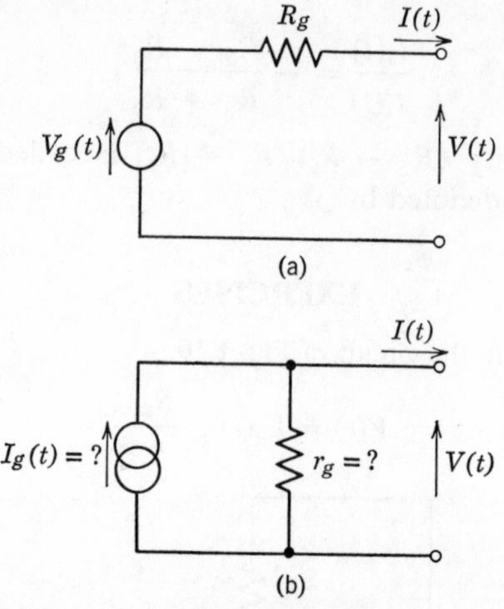

Figure 1.12

4. Find V_g and R_g in the circuit of Fig. 1.13(b) ("Thevenin equivalent circuit") so that it is equivalent to the bridge circuit of Fig. 1.13(a) at the terminals described by V and I. Evaluate V_g and R_g for $R_1 = 0.5\ \Omega$, $1\ \Omega$, and $2\ \Omega$.

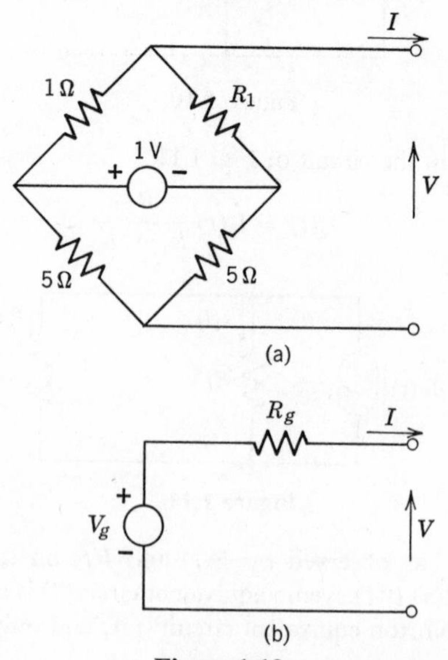

Figure 1.13

5. Derive Equation (1.6) by using Equation (1.3) and the energy relationship

$$E(t) = \int_{-\infty}^{t} V(t')I(t')\,dt'.$$

6. Derive Equation (1.10) by using Equation (1.9) and the energy relationship

$$E(t) = \int_{-\infty}^{t} V(t')I(t')\,dt'.$$

7. The amount of charge $Q(t)$ transferred in a circuit is the integral of the current:

$$Q(t) = \int I(t)\,dt.$$

The unit of the charge is the coulomb: 1 coulomb = 1 ampere flowing for 1 second. Sketch the charge $Q(t)$ on the capacitor and the terminal voltage $V(t)$ in the circuit of Fig. 1.14 with $V(t < 0) = 0$.

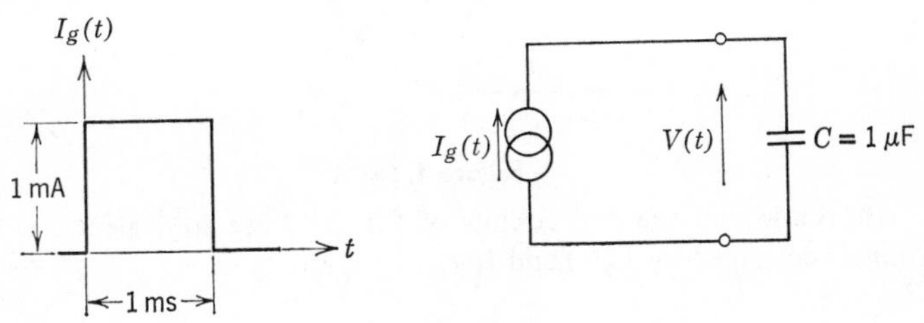

Figure 1.14

8. The magnetic flux $\Phi(t)$ is the integral of the voltage: $\Phi(t) = \int V(t)\,dt$, where $V(t)$ is the voltage per turn. The unit of the flux is the weber: 1 weber = 1 volt present for 1 second. Sketch the current $I(t)$ and the flux $\Phi(t)$ of the inductor in the circuit of Fig. 1.15 with $I(t < 0) = 0$.

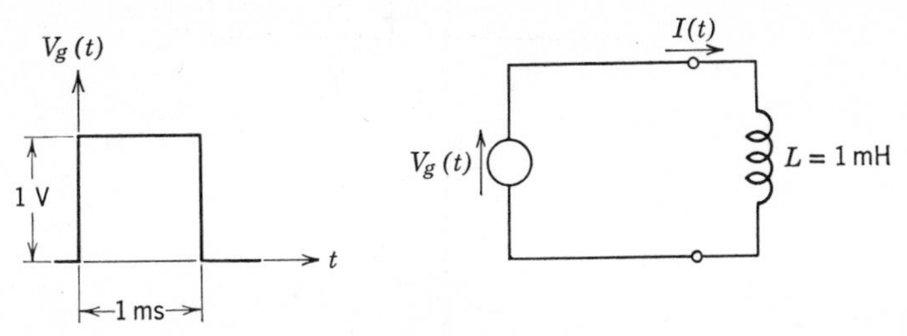

Figure 1.15

9. Show that in the circuit of Fig. 1.16(a) utilizing an ideal transformer,

$$\frac{V_1(t)}{I_1(t)} = R_T/N^2 \quad \text{as shown in Fig. 1.16(b).}$$

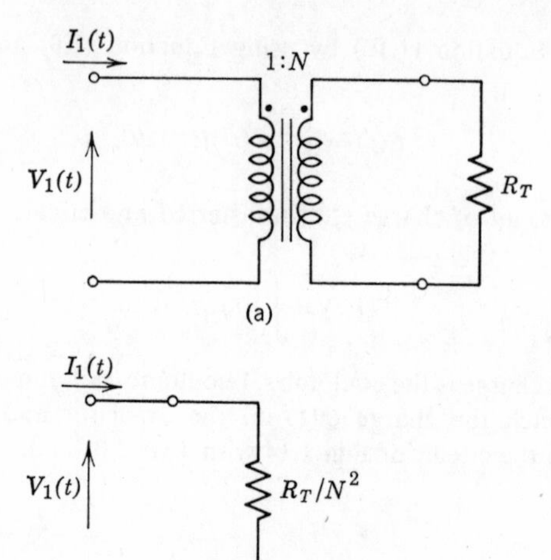

Figure 1.16

10. Show that the two circuits of Fig. 1.17 are equivalent on their terminals described by $V_1(t)$ and $I_1(t)$.

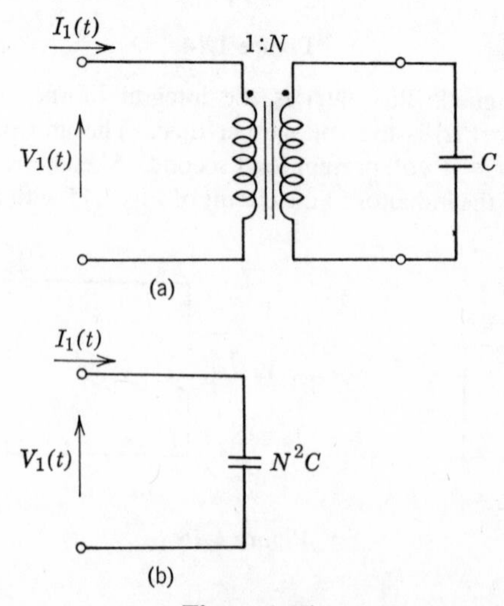

Figure 1.17

11. Show that the two circuits of Fig. 1.18 are equivalent on their terminals described by $V_1(t)$ and $I_1(t)$.

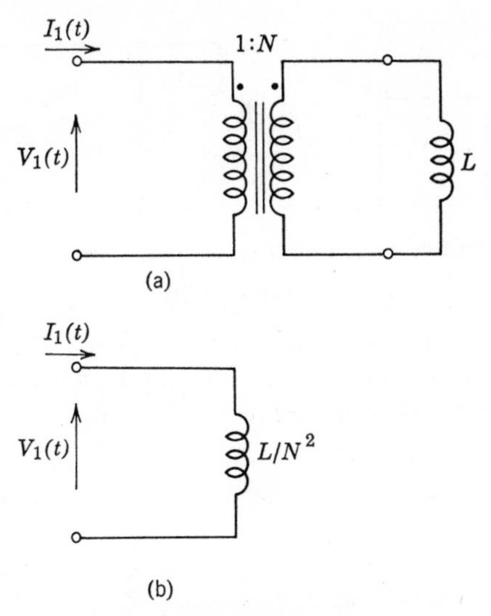

Figure 1.18

12. Show that in the circuit of Fig. 1.19(a), ($R_T = 0$), the voltage $V_1(t)$ is as shown in Fig. 1.19(b) if $V_1(t < 0) = 0$.

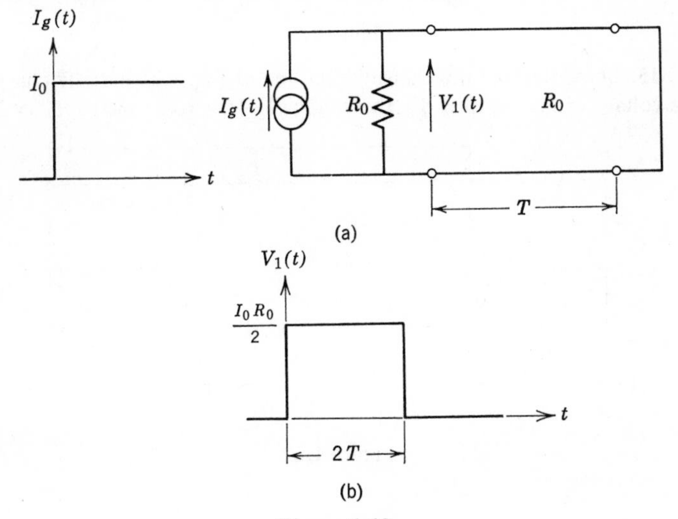

Figure 1.19

13. Sketch the voltage waveform $V_1(t)$ in the circuit of Fig. 1.20 for $T = 5$ ns, $T = 10$ ns, and $T = 20$ ns, assuming $V_1(t < 0) = 0$.

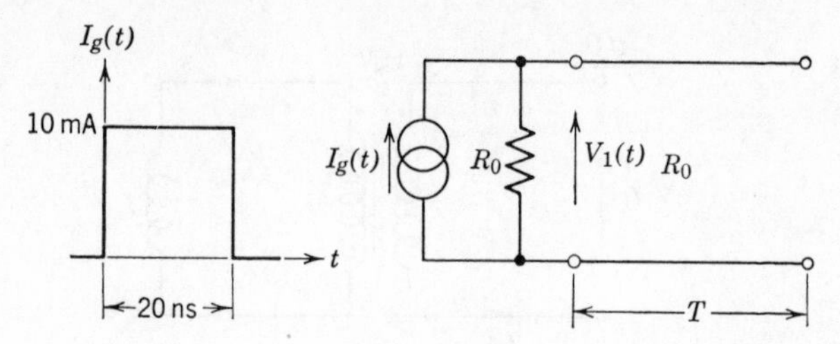

Figure 1.20

14. Show that for the transmission line of Fig. 1.21(a) initially charged to a voltage of $V_1(t < 0) = V_0$, the resulting $V_2(t)$ is that of Fig. 1.21(b), if switch S is closed at time $t = 0$ and if $R_g \gg R_0$.

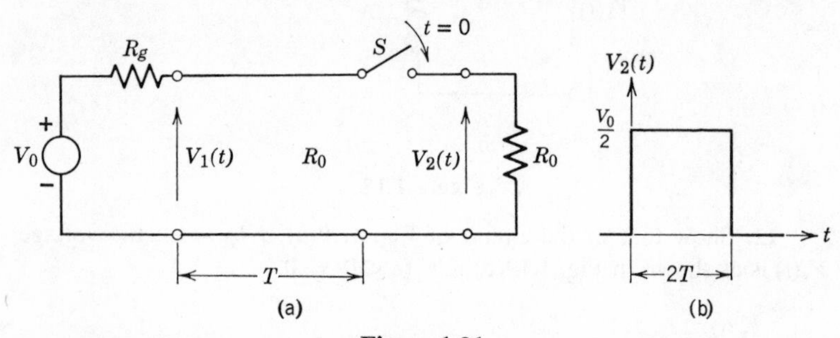

Figure 1.21

15. Show that for the transmission line of Fig. 1.22 initially charged to a voltage of $V_2(t < 0) = V_0$, the voltage $V_L(t)$ across resistor $R_L = 2R_0$

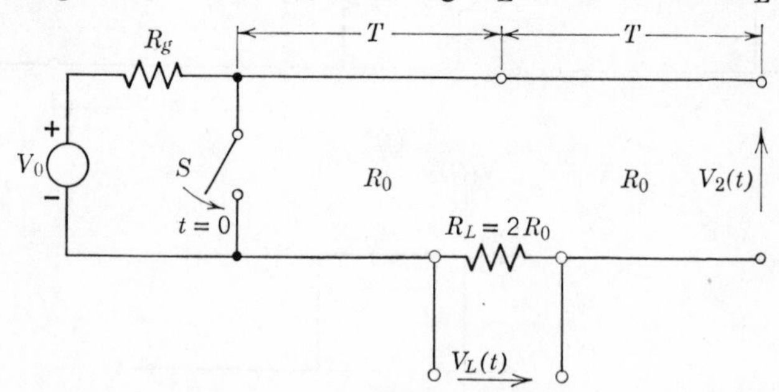

Figure 1.22

is a single pulse of length $2T$ and height V_0 if ideal switch S is closed at time $t = 0$ and if $R_g \gg R_0$. Ignore the length of resistor R_L.

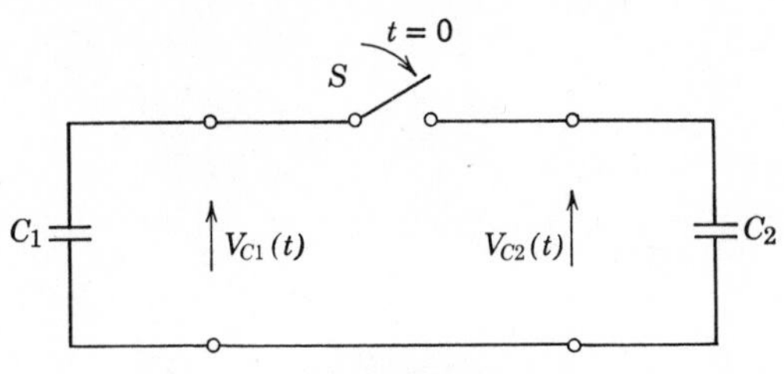

Figure 1.23

16. In Fig. 1.4, the load of an ideal voltage source is restricted such that the terminal current is finite for a finite source voltage. Use this restriction to show that the closure of ideal switch S in Fig. 1.23 is not allowed unless $V_{C1}(t < 0) = V_{C2}(t < 0)$.

OOOOO

2

OOOOO

The Laplace Transform and its Application to Transients in Linear Circuits

OOOOOOOOOOOOOOOOOOOOOOOOOOOOOOO

The Laplace transform is a mathematical aid to the solution of differential equations arising in the transient analysis of linear circuits. If $f(t)$ is a real function such that $f(t < 0) = 0$, then its Laplace transform, denoted by $\mathscr{L}\{f(t)\}$, or by $F(s)$, is defined as

$$F(s) \equiv \mathscr{L}\{f(t)\} \equiv \int_0^\infty f(t)e^{-st}\,dt. \tag{2.1}$$

In addition to requiring $f(t)$ to be real, the following additional restrictions will be made: There exist positive constants M, σ, and t_0 such that

$$|f(t)| \leq Me^{\sigma t} \quad \text{for every} \quad t \geq t_0. \tag{2.2}$$

With these restrictions, the integral of Equation (2.1) converges absolutely for all s with real part $\text{Re}(s) > \sigma$. (See Apostol in references.) In what follows, the restriction $f(t < 0) = 0$, which is inherent in the definition of Equation (2.1), will be always observed, although not always emphasized.

LAPLACE TRANSFORMS OF SOME SIMPLE FUNCTIONS

1. *Unit step:*

$$f(t) = u(t) \tag{2.3a}$$

where

$$u(t) = \begin{cases} 0, & t < 0 \\ 1, & t > 0. \end{cases} \tag{2.3b}$$

Utilizing Equation (2.1),

$$\mathscr{L}\{u(t)\} = \int_0^\infty e^{-st}\,dt = \frac{1}{s}. \tag{2.4}$$

2. *Linear ramp:*

$$f(t) = t,\ t > 0;\ f(t < 0) = 0:$$

$$\mathscr{L}\{t\} = \int_0^\infty te^{-st}\,dt = \frac{1}{s^2}. \tag{2.5}$$

3. *Exponential:*

$$f(t) = e^{-\alpha t},\ t > 0;\ f(t < 0) = 0:$$

$$\mathscr{L}\{e^{-\alpha t}\} = \int_0^\infty e^{-\alpha t}e^{-st}\,dt = \frac{1}{s + \alpha}. \tag{2.6}$$

4. *Sinewave:*

$$f(t) = \sin \alpha t,\ t > 0;\ f(t < 0) = 0:$$

$$\mathscr{L}\{\sin \alpha t\} = \frac{\alpha}{s^2 + \alpha^2}. \tag{2.7}$$

Additional transforms are listed in Table 2.1.

BASIC LAPLACE TRANSFORM THEOREMS

From the definition of Equation (2.1) follows the important property of linearity that can be expressed as $\mathscr{L}\{f_1(t) + f_2(t)\} = \mathscr{L}\{f_1(t)\} + \mathscr{L}\{f_2(t)\}$, or as $\mathscr{L}\{Kf(t)\} = K\mathscr{L}\{f(t)\}$ where K is a constant. Several additional properties of the Laplace transform will be given here as theorems without proof; in all cases $F(s)$ is defined as $\mathscr{L}\{f(t)\}$.

1. *Shift theorem:*

$$\mathscr{L}\{f(t - T)u(t - T)\} = e^{-sT}F(s). \tag{2.8}$$

2. *Scale-change theorem:*

$$\mathscr{L}\left\{f\left(\frac{t}{T}\right)\right\} = \int_0^\infty f\left(\frac{t}{T}\right) e^{-st}\, dt = TF(Ts). \tag{2.9}$$

3. *Convolution theorem:*

$$\mathscr{L}\left\{\int_0^t f_1(\xi) f_2(t-\xi)\, d\xi\right\} = \mathscr{L}\{f_1(t)\}\mathscr{L}\{f_2(t)\}. \tag{2.10}$$

4. *Differentiation theorem:*

$$\mathscr{L}\left\{\frac{df(t)}{dt}\right\} = sF(s) - f(t = 0^+). \tag{2.11}$$

5. *Integration theorem:*

$$\mathscr{L}\left\{\int_0^t f(t')\, dt'\right\} = \frac{F(s)}{s} + \frac{1}{s}\int_0^{t'=0^+} f(t')\, dt'. \tag{2.12}$$

6. *Initial value theorem:*

$$\lim_{t\to 0} f(t) = \lim_{s\to\infty} sF(s). \tag{2.13}$$

7. *Final value theorem:*

$$\lim_{t\to\infty} f(t) = \lim_{s\to 0} sF(s). \tag{2.14}$$

INVERSION OF THE LAPLACE TRANSFORM

In general, when $F(s) \equiv \mathscr{L}\{f(t)\}$ is given, $f(t) = \mathscr{L}^{-1}\{F(s)\}$ can be determined by using the inversion formula

$$f(t) = \frac{1}{2\pi j}\int_{\sigma_1 - j\omega}^{\sigma_1 + j\omega} F(s)\, ds, \tag{2.15}$$

where $\sigma_1 < \sigma$ of Equation (2.2). In circuit analysis, however, this inversion formula is rarely used since extensive tables of Laplace transform pairs are available. (See Erdelyi, Nixon, and Roberts in references.) A short table is given in Table 2.1. When a function $F(s)$ is to be inverted, an attempt should be made to convert it to one available in the tables, and Equation (2.15) should be used only if the tables can not be utilized.

Table 2.1 A short table of Laplace-Transform pairs.

$f(t)$ for $t > 0$	$F(s) \equiv \mathscr{L}\{f(t)\}$	Conditions
$\delta(t-a)$	e^{-as}	$a \geq 0$; $\mathrm{Re}(s) > -\infty$
1	$\frac{1}{s}$	$\mathrm{Re}(s) > 0$
e^{-at}	$\frac{1}{s+a}$	$\mathrm{Re}(s) > -\mathrm{Re}(a)$
te^{-at}	$\frac{1}{(s+a)^2}$	$\mathrm{Re}(s) > -\mathrm{Re}(a)$
$e^{-bt} - e^{-at}$	$\frac{a-b}{(s+a)(s+b)}$	$\mathrm{Re}(s) > \mathrm{Max}[-\mathrm{Re}(a), -\mathrm{Re}(b)]$; $a \neq b$
$ae^{-at} - be^{-bt}$	$\frac{(a-b)s}{(s+a)(s+b)}$	$\mathrm{Re}(s) > \mathrm{Max}[-\mathrm{Re}(a), -\mathrm{Re}(b)]$; $a \neq b$
$1 - e^{-at}$	$\frac{a}{s(s+a)}$	$\mathrm{Re}(s) > \mathrm{Max}[0, -\mathrm{Re}(a)]$

$(1 - at)e^{-at}$	$\frac{s}{(s + a)^2}$	$\mathrm{Re}(s) > \mathrm{Max}[0, -\mathrm{Re}(a)]$
$\frac{t^{n-1}e^{-at}}{(n - 1)!}$	$\frac{1}{(s + a)^n}$	$n = 1, 2, 3, \ldots;\ \mathrm{Re}(s) > \mathrm{Max}[0, -\mathrm{Re}(a)]$
$1 - e^{-at}\left(1 + \frac{at}{1!} + \cdots + \frac{(at)^n}{n!}\right)$	$\frac{a^{n+1}}{s(s + a)^{n+1}}$	$n = 0, 1, 2, \ldots;\ \mathrm{Re}(s) > \mathrm{Max}[0, -\mathrm{Re}(a)]$
$a - b - ae^{-bt} + be^{-at}$	$\frac{ab(a - b)}{s(s + a)(s + b)}$	$\mathrm{Re}(s) > \mathrm{Max}[-\mathrm{Re}(a), -\mathrm{Re}(b)]$
$\sin(at)$	$\frac{a}{s^2 + a^2}$	$\mathrm{Re}(s) > \lvert\mathrm{Im}(a)\rvert$
$\cos(at)$	$\frac{s}{s^2 + a^2}$	$\mathrm{Re}(s) > \lvert\mathrm{Im}(a)\rvert$
$\frac{1 - a^n}{1 - a}$	$\frac{1}{s(e^{bs} - a)}$	$nb < t < (n + 1)b;\ n = 0, 1, \ldots;$ $\mathrm{Re}(s) > 0,\ b\,\mathrm{Re}(s) > \mathrm{Re}[\ln(a)]$
$\frac{a}{\sqrt{\pi}} e^{-a^2(t-D)^2}$	$e^{(s/2a)^2} e^{-sD}$	$a > 0;\ \mathrm{Re}(s) > -\infty;\ e^{a^2D^2} \gg 1$

TRANSIENT RESPONSE OF AN *R-C* CIRCUIT

As an application of the Laplace transform technique, the transient response of the *R-C* circuit of Fig. 2.1 will be analyzed with $V_C(t < 0) = 0$. The following equations can be written:

$$V_g(t) = V_R(t) + V_C(t) \tag{2.16}$$

$$V_R(t) = RI(t) \tag{2.17}$$

$$I(t) = C\frac{dV_C(t)}{dt}, \tag{2.18}$$

which can be combined as

$$V_g(t) = RC\frac{dV_C(t)}{dt} + V_C(t). \tag{2.19}$$

Step-Function Input Voltage. In this case, $V_g(t) = V_0u(t)$, i.e.,

$$V_g(t) = \begin{cases} 0, & t < 0 \\ V_0, & t > 0 \end{cases},$$

hence $\mathscr{L}\{V_g(t)\} = V_0/s$. The Laplace transform of Equation (2.19) becomes

$$\frac{V_0}{s} = RC[s\mathscr{L}\{V_C(t)\} - V_C(t = 0^+)] + \mathscr{L}\{V_C(t)\}. \tag{2.20}$$

The value of $V_C(t < 0)$ is zero. A finite voltage V_0 at $t = 0$ results in a finite current $I(t)$ and thus in a finite dV_C/dt [See Equation

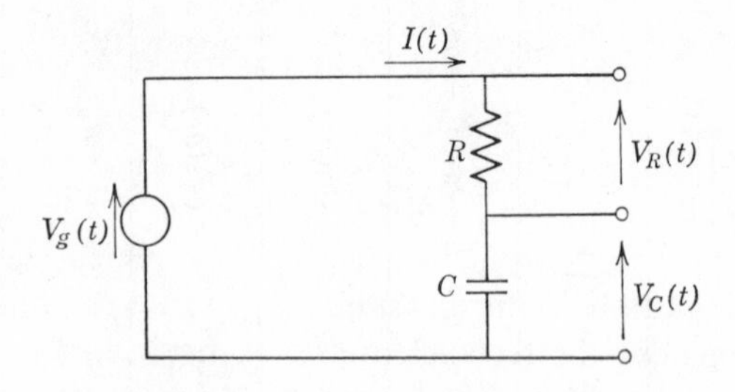

Figure 2.1 A series *R-C* circuit.

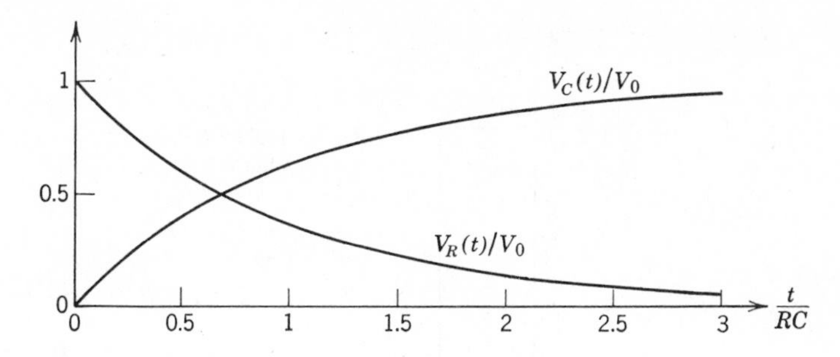

Figure 2.2 Transient response of the circuit of Fig. 2.1 for a step-function input voltage.

(2.18)]. For this reason, $V_C(t = 0^+) = V_C(t = 0^-) = 0$. From Equation (2.20),

$$\mathscr{L}\{V_C(t)\} = \frac{V_0}{RC}\frac{1}{s(s + 1/RC)}\,. \tag{2.21}$$

The inverse Laplace transform from Table 2.1 is

$$V_C(t) = V_0(1 - \mathrm{e}^{-t/RC}), \tag{2.22}$$

and the voltage on the resistor, by using Equation (2.16), is

$$V_R(t) = V_0\mathrm{e}^{-t/RC}; \tag{2.23}$$

these signals are shown in Fig. 2.2.

δ-Function Input Voltage. The unit δ-function (delta-function), or impulse function, $\delta(t)$, has the properties (See Bracewell in references)

$$\delta(t) = 0 \quad \text{for} \quad t \neq 0 \tag{2.24}$$

and

$$\int_{-\infty}^{\infty} \delta(t)\, dt = 1. \tag{2.25}$$

The δ-function can be defined as a limiting case of several functions; one such function, the rectangle-pulse, is shown in Fig. 2.3. It is seen from the definition that for this function in the limit $T \rightarrow 0$ both Equations (2.24) and (2.25) are valid.

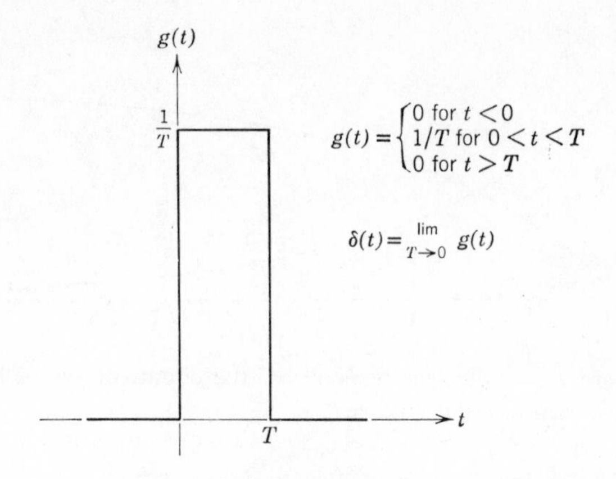

Figure 2.3 The δ-function defined as the limit of the rectangle-pulse.

An inspection of the unit step-function $u(t)$ shows a relationship between it and the unit δ-function:

$$\int_{-\infty}^{t} \delta(t')\, dt' = u(t), \tag{2.26}$$

which also implies

$$\delta(t) = \frac{du(t)}{dt}. \tag{2.27}$$

Since the derivative of $u(t)$ does not exist at $t = 0$, Equation (2.27) is not rigorous, and the extent of its validity is questionable. It can be shown, however, that as far as Laplace transforms are concerned, Equation (2.27) is applicable.

In analyzing the transient response for a δ-function input, three methods may be used. The first of these is based on Equation (2.27): The response for a $V_g(t) = \delta(t)$ is the derivative of the response for $V_g(t) = u(t)$. For a $V_g(t) = \Phi_0\delta(t)$, by using the derivatives of Equations (2.22) and (2.23), for $t > 0$,

$$V_C(t) = \frac{\Phi_0}{RC}\, e^{-t/RC} \tag{2.28}$$

and

$$V_R(t) = -\frac{\Phi_0}{RC}\, e^{-t/RC}. \tag{2.29}$$

In the second method, these results can be derived by taking the δ-function voltage source into account as an initial condition: Immediately after the δ-function returns to zero ($t = 0^+$), the voltage on the capacitor is

$$V_C(t = 0^+) = \frac{1}{C}\int I(t)\,dt = \frac{\Phi_0}{RC}. \tag{2.30}$$

Substituting

$$\mathscr{L}\left\{\frac{dV_C(t)}{dt}\right\} = s\mathscr{L}\{V_C(t)\} - \frac{\Phi_0}{RC} \tag{2.31}$$

and $V_g(t) = 0$ into Equation (2.19) and utilizing Equation (2.16) again result in Equations (2.28) and (2.29).

The third approach enters the δ-function at a time $t > 0^+$. Thus, $V_C(t = 0^+) = 0$ and $\mathscr{L}\{V_g(t)\} = \mathscr{L}\{\Phi_0\delta(t)\} = \Phi_0$, leading to the same results as above.

TRANSIENT RESPONSE OF A SERIES *R-L-C* CIRCUIT

As a further example of Laplace transform techniques, the current $I(t)$ of the series *R-L-C* circuit of Fig. 2.4 will be computed for a step-function voltage input and for zero initial conditions. The following equations can be written:

$$V_g(t) = V_R(t) + V_L(t) + V_C(t) \tag{2.32}$$

$$I(t) = V_R(t)/R \tag{2.33}$$

$$I(t) = \frac{1}{L}\int_0^t V_L(t')\,dt' \tag{2.34}$$

$$I(t) = C\,dV_C(t)/dt. \tag{2.35}$$

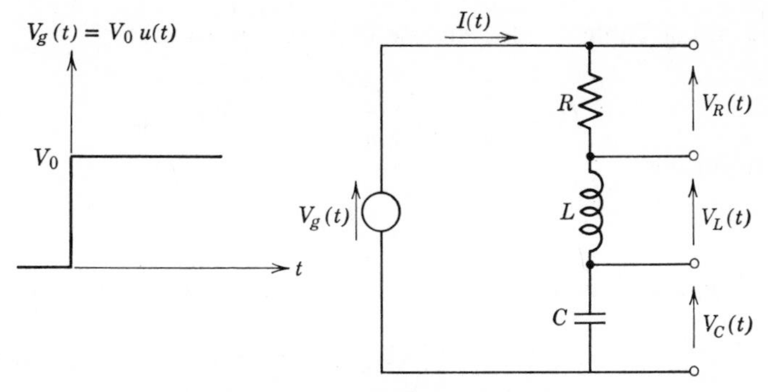

Figure 2.4 A series *R-L-C* circuit.

For a finite $V_g(t)$,

$$\mathscr{L}\left\{\frac{dV_C(t)}{dt}\right\} = s\mathscr{L}\{V_C(t)\} - V_C(t = 0^+)$$
$$= s\mathscr{L}\{V_C(t)\} - V_C(t = 0^-) = s\mathscr{L}\{V_C(t)\} \qquad (2.36)$$

and

$$\mathscr{L}\left\{\int_0^t V_L(t)\,dt\right\} = \frac{1}{s}\mathscr{L}\{V_L(t)\} + \frac{1}{s}\int_0^{t=0^+} V_L(t')\,dt' = \frac{1}{s}\mathscr{L}\{V_L(t)\}. \qquad (2.37)$$

Combining Equations (2.32) through (2.37) and substituting $\mathscr{L}\{V_g(t)\} = V_0/s$ results in

$$\mathscr{L}\{I(t)\} = \frac{V_0}{L}\frac{1}{s^2 + \frac{R}{L}s + \frac{1}{LC}}. \qquad (2.38)$$

The roots of the denominator are:

$$a_1 = -\frac{R}{2L} - \sqrt{\left(\frac{R}{2L}\right)^2 - \frac{1}{LC}} = -\frac{R}{2L}\left[1 + \sqrt{1 - \frac{4L}{R^2C}}\right] \qquad (2.39)$$

and

$$a_2 = -\frac{R}{2L} + \sqrt{\left(\frac{R}{2L}\right)^2 - \frac{1}{LC}} = -\frac{R}{2L}\left[1 - \sqrt{1 - \frac{4L}{R^2C}}\right], \qquad (2.40)$$

hence Equation (2.38) can be written as

$$\mathscr{L}\{I(t)\} = \frac{V_0}{L}\frac{1}{(s - a_1)(s - a_2)}, \qquad (2.41)$$

and, using Table 2.1, the inverse Laplace transform is

$$I(t) = \frac{V_0}{L}\frac{e^{a_2 t} - e^{a_1 t}}{a_2 - a_1}; \qquad a_1 \neq a_2. \qquad (2.42)$$

Introducing

$$m \equiv \frac{L}{R^2C}, \qquad (2.43)$$

roots a_1 and a_2 become

$$a_1 = -\frac{R}{2L}(1 + \sqrt{1 - 4m}) \qquad (2.44)$$

and

$$a_2 = -\frac{R}{2L}(1 - \sqrt{1 - 4m}). \tag{2.45}$$

Depending on the value of m, there are three possible forms of solution:

1. Case of two distinct real roots ($m < 0.25$)—*Overdamped.* Defining

$$D^2 \equiv \left(\frac{R}{2L}\right)^2 - \frac{1}{LC} = \left(\frac{R}{2L}\right)^2(1 - 4m), \tag{2.46}$$

the roots become

$$a_1 = -\frac{R}{2L} - D$$

and

$$a_2 = -\frac{R}{2L} + D.$$

The current $I(t)$ of Equation (2.42) can now be written in several forms:

$$I(t) = \frac{V_0}{2LD}(e^{a_2 t} - e^{a_1 t}), \tag{2.47a}$$

or

$$I(t) = \frac{V_0}{2LD}\,e^{(-R/2L+D)t}(1 - e^{-2Dt}), \tag{2.47b}$$

or

$$I(t) = \frac{V_0}{LD}\,e^{-(R/2L)t}\sinh Dt. \tag{2.47c}$$

The first form shows a direct solution assembled from the two roots. The second form is useful in evaluating $I(t)$ for large values of t, the third form for small values of t.

2. Case of two equal real roots ($m = 0.25$)—*Critically damped.* In this case, $a_1 = a_2 = a = -R/2L$ and Equation (2.41) becomes

$$\mathscr{L}\{I(t)\} = \frac{V_0}{L}\frac{1}{(s - a)^2}. \tag{2.48}$$

Utilizing the inverse transform from Table 2.1,

$$I(t) = \frac{V_0}{L}\,te^{at} = \frac{V_0}{L}\,te^{-(R/2L)t}. \tag{2.49}$$

3. Case of unequal complex roots ($m > 0.25$)—*Underdamped* or *Oscillatory*. Defining a ringing frequency ω_0 by

$$\omega_0{}^2 = \frac{1}{LC}\left(1 - \frac{R^2C}{4L}\right) = \frac{1}{LC}\left(1 - \frac{1}{4m}\right),$$

the roots become

$$a_1 = -\frac{R}{2L} - j\omega_0$$

and

$$a_2 = -\frac{R}{2L} + j\omega_0,$$

and Equation (2.42) becomes

$$I(t) = \frac{V_0}{L}\,\frac{\exp\left[\left(-\dfrac{R}{2L} + j\omega_0\right)t\right] - \exp\left[\left(-\dfrac{R}{2L} - j\omega_0\right)t\right]}{2j\omega_0}, \tag{2.50a}$$

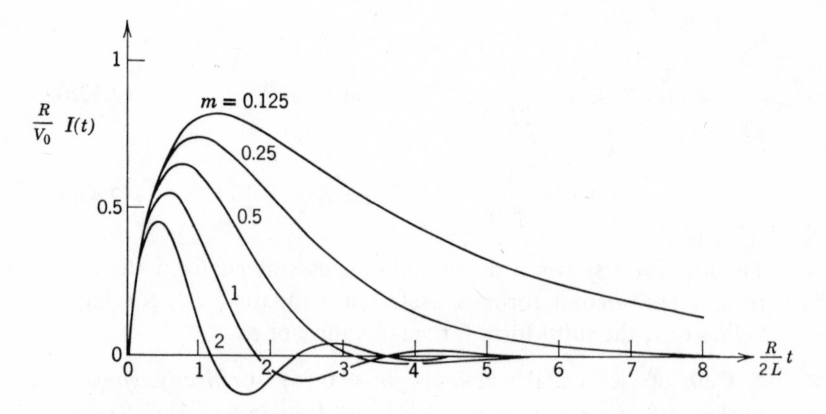

Figure 2.5 Transient response of the circuit of Fig. 2.4 for $m = 0.125$ (*overdamped*); for $m = 0.25$ (*critically damped*); and for $m = 0.5$, 1, and 2 (*underdamped*).

which can also be written as

$$I(t) = \frac{V_0}{\omega_0 L} e^{-(R/2L)t} \sin \omega_0 t. \tag{2.50b}$$

Waveforms corresponding to the three cases are shown in Fig. 2.5; as expected, they are oscillatory for $m > 0.25$.

TRANSIENT RESPONSE OF A SERIES-PARALLEL *R-L-C* CIRCUIT

The transient response of the series-parallel *R-L-C* circuit of Fig. 2.6 with a step-function input current and zero initial conditions will be summarized. For this circuit,

$$\mathscr{L}\{V(t)\} = \frac{I_0}{C}\left[\frac{1}{(s - a_1)(s - a_2)} + \frac{R}{L}\frac{1}{s(s - a_1)(s - a_2)}\right], \tag{2.51}$$

where

$$a_1 = -\frac{R}{2L}\left[1 + \sqrt{1 - \frac{4L}{R^2C}}\right]$$

and

$$a_2 = -\frac{R}{2L}\left[1 - \sqrt{1 - \frac{4L}{R^2C}}\right].$$

If $m \equiv L/R^2C < 0.25$ (*overdamped* case), then

$$V(t) = I_0 R\left\{1 - e^{-(R/2L)t}\left[\cosh Dt + \frac{R}{2LD}\left(1 - \frac{2L}{R^2C}\right) \sinh Dt\right]\right\}, \tag{2.52}$$

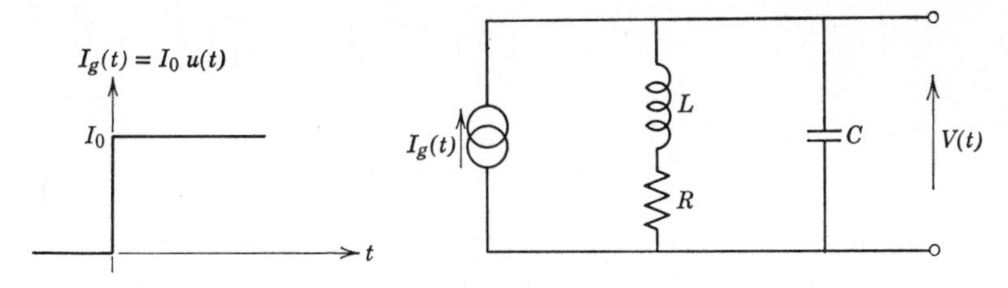

Figure 2.6 A series-parallel *R-L-C* circuit.

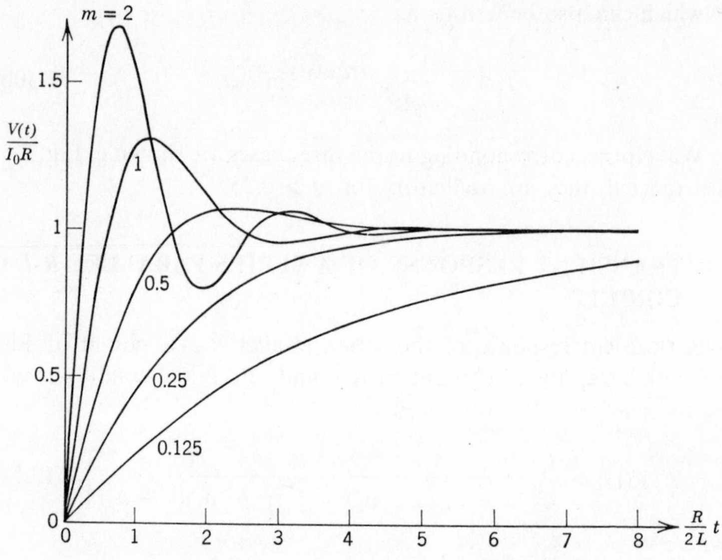

Figure 2.7 Transient response of the circuit of Fig. 2.6 for $m = 0.125$ (*overdamped*); for $m = 0.25$ (*critically damped*); and for $m = 0.5$, 1, and 2 (*underdamped*).

where

$$D^2 \equiv \left(\frac{R}{2L}\right)^2 - \frac{1}{LC} = \left(\frac{R}{2L}\right)^2(1 - 4m).$$

If $m \equiv L/R^2C = 0.25$ (*critically damped* case), then

$$V(t) = I_0 R\left[1 - \frac{R}{4L}\, t e^{-(R/2L)t} - e^{-(R/2L)t}\right]. \tag{2.53}$$

If $m \equiv L/R^2C > 0.25$ (*underdamped* case), then

$$V(t) = I_0 R\left\{1 - e^{-(R/2L)t}\left[\cos\omega_0 t + \frac{R}{2L\omega_0}\left(1 - \frac{2L}{R^2C}\right)\sin\omega_0 t\right]\right\} \tag{2.54}$$

where

$$\omega_0{}^2 \equiv \frac{1}{LC}\left(1 - \frac{R^2C}{4L}\right) = \frac{1}{LC}\left(1 - \frac{1}{4m}\right).$$

Normalized voltage $V(t)/I_0R$ as function of $(R/2L)t$ is shown in Fig. 2.7 for various values of m. It is seen that for $m > 0.25$ the response is oscillatory, as expected.

TRANSIENTS IN IDEAL TRANSMISSION LINES

The Laplace transform can also be used to analyze transients on ideal transmission lines. Consider the ideal transmission line of Fig. 2.8, driven by a voltage source with a source resistance R_g and terminated by a resistance R_L. The output signal $V_2(t)$ of this circuit will be computed for an input step function $V_g(t)$, with the line initially uncharged, i.e., with $V_1(t < 0) = V_2(t < 0) = 0$.

The voltages $V_1(t)$ and $V_2(t)$ can be expressed as the sum of incident and reflected signals:

$$V_1(t) = V_{i1}(t) + V_{r1}(t), \tag{2.55}$$

$$V_2(t) = V_{i2}(t) + V_{r2}(t). \tag{2.56}$$

Similarly, $I_1(t)$ and $I_2(t)$ can be expressed as

$$I_1(t) = I_{i1}(t) + I_{r1}(t), \tag{2.57}$$

$$I_2(t) = I_{i2}(t) + I_{r2}(t). \tag{2.58}$$

The current $I_1(t)$ can also be written as

$$I_1(t) = \frac{V_g(t) - V_1(t)}{R_g}, \tag{2.59}$$

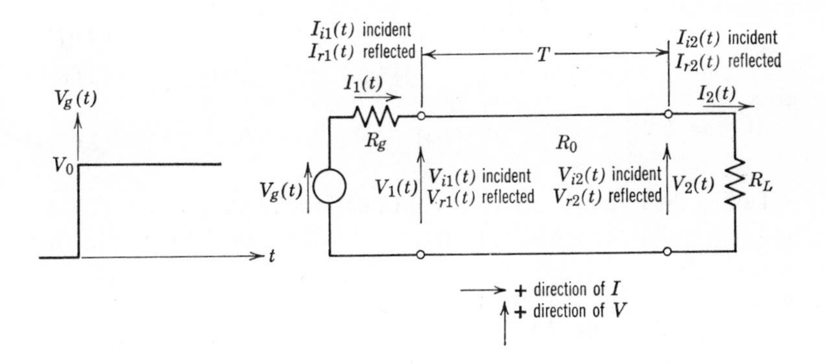

Figure 2.8 Transmission line driven by a source with an impedance of R_g and terminated by an impedance of R_L.

and $I_2(t)$ can be written as

$$I_2(t) = \frac{V_2(t)}{R_L}. \tag{2.60}$$

On an ideal transmission line the traveling voltage and current are related by R_0. With the positive directions of I and V as chosen,

$$\frac{V_{i1}}{I_{i1}} = -R_0, \tag{2.61}$$

$$\frac{V_{r1}}{I_{r1}} = R_0, \tag{2.62}$$

$$\frac{V_{i2}}{I_{i2}} = R_0, \tag{2.63}$$

and

$$\frac{V_{r2}}{I_{r2}} = -R_0. \tag{2.64}$$

An ideal transmission line delays signals with no distortion. Therefore,

$$V_{i1}(t) = V_{r2}(t - T) \tag{2.65}$$

and

$$V_{i2}(t) = V_{r1}(t - T). \tag{2.66}$$

Since all signals are zero for negative times, the Laplace transforms of Equations (2.65) and (2.66) can be written as

$$\mathscr{L}\{V_{r2}(t - T)u(t - T)\} = e^{-sT}\mathscr{L}\{V_{r2}(t)\}, \tag{2.67}$$

$$\mathscr{L}\{V_{r1}(t - T)u(t - T)\} = e^{-sT}\mathscr{L}\{V_{r1}(t)\}; \tag{2.68}$$

also,

$$\mathscr{L}\{V_g(t)\} = \frac{V_0}{s}. \tag{2.69}$$

Taking the Laplace transforms of Equations (2.55) through (2.64) and combining them with Equations (2.67) through (2.69) results in

$$\mathscr{L}\{V_2(t)\} = \frac{2V_0R_L/R_0}{(1 + R_g/R_0)(1 + R_L/R_0)}\,\frac{e^{sT}}{s} \times \frac{1}{e^{2sT} - \dfrac{(1 - R_g/R_0)(1 - R_L/R_0)}{(1 + R_g/R_0)(1 + R_L/R_0)}}. \tag{2.70}$$

To find the inverse Laplace transform, consider from Table 2.1:

$$\mathscr{L}^{-1}\left\{\frac{1}{s}\frac{1}{e^{bs}-a}\right\} = \frac{1-a^n}{1-a}, \qquad nb < t < (n+1)b,$$
$$n = 0, 1, 2, \ldots.$$

Therefore, by substituting and simplifying,

$$V_2(t) = V_0\frac{R_L}{R_g + R_L}\left\{1 - \left[\frac{(1 - R_g/R_0)(1 - R_L/R_0)}{(1 + R_g/R_0)(1 + R_L/R_0)}\right]^N\right\}, \qquad (2.71a)$$

where the range of t for each N is defined as:

$$2T(N - \tfrac{1}{2}) < t < 2T(N + \tfrac{1}{2}); \qquad t > 0. \qquad (2.71b)$$

Thus $V_2(t)$ is composed of a series of steps; $V_2(t)/V_0$ is shown in Fig. 2.9 for various values of R_g and R_L.

If $R_g \gg R_0$ and $R_L \gg R_0$, an approximate exponential form is possible. Equation (2.71a) can be written as

$$V_2(t) = V_0\frac{R_L}{R_g + R_L}$$
$$\times \left\{1 - \exp\left[N \ln\left(1 - 2\frac{R_g/R_0 + R_L/R_0}{(1 + R_g/R_0)(1 + R_L/R_0)}\right)\right]\right\}, \qquad (2.72)$$

with the range of t as given by Equation (2.71b).

If $R_g \gg R_0$, and $R_L \gg R_0$, then, since $\ln(1 + x) \approx x$ for $|x| \ll 1$,

$$V_2(t) \approx V_0\frac{R_L}{R_g + R_L}\left\{1 - \exp\left[-2N\frac{R_g/R_0 + R_L/R_0}{(R_g/R_0)(R_L/R_0)}\right]\right\}, \qquad (2.73)$$

again with the same range of t as in Equation (2.7lb). When $t = 2TN$, Equation (2.73) becomes

$$V_2(t) \approx V_0\frac{R_L}{R_g + R_L}\left\{1 - \exp\left[-\frac{t}{T/R_0}\frac{R_g + R_L}{R_gR_L}\right]\right\}. \qquad (2.74)$$

This, however, is the same transient as if the transmission line were replaced by a capacitance of $C = T/R_0$.* Thus, for $R_g \gg R_0$ and

* This capacitance, known as the capacitance of the transmission line, is given as such in transmission line tables.

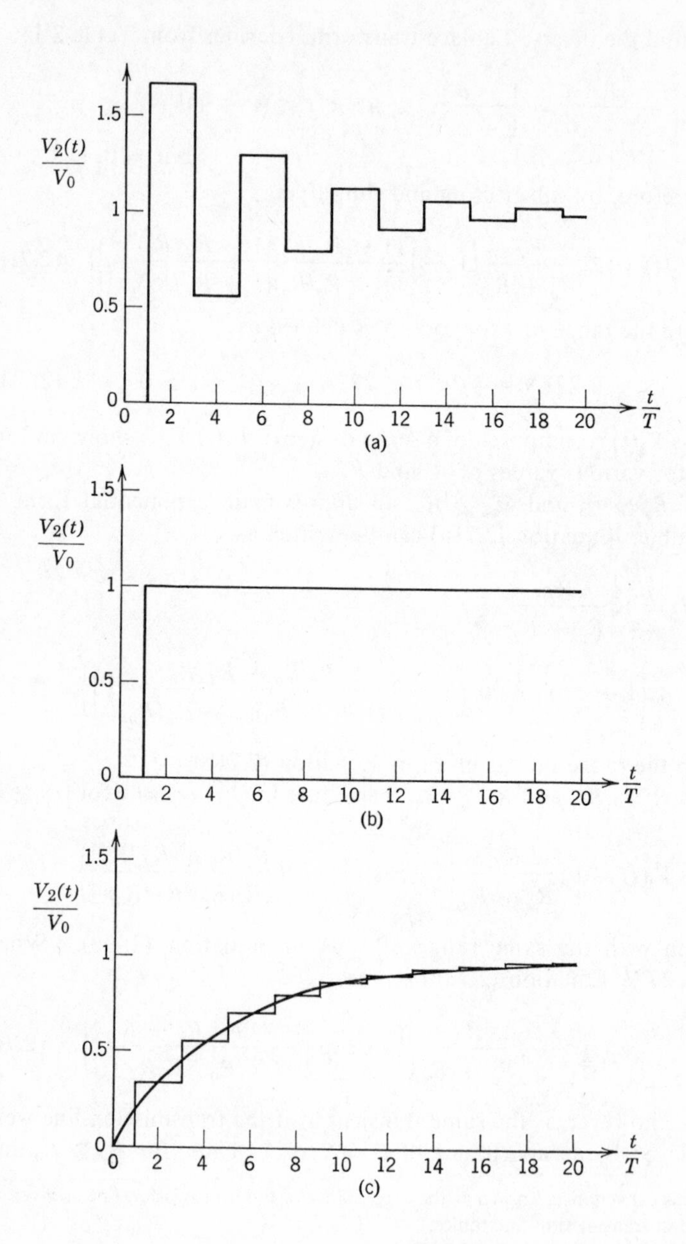
1.5
1
0.5
0
$\frac{V_2(t)}{V_0}$
0 2 4 6 8 10 12 14 16 18 20
$\frac{t}{T}$
(a)
1.5
1
0.5
0
$\frac{V_2(t)}{V_0}$
0 2 4 6 8 10 12 14 16 18 20
$\frac{t}{T}$
(b)
1.5
1
0.5
0
$\frac{V_2(t)}{V_0}$
0 2 4 6 8 10 12 14 16 18 20
$\frac{t}{T}$
(c)

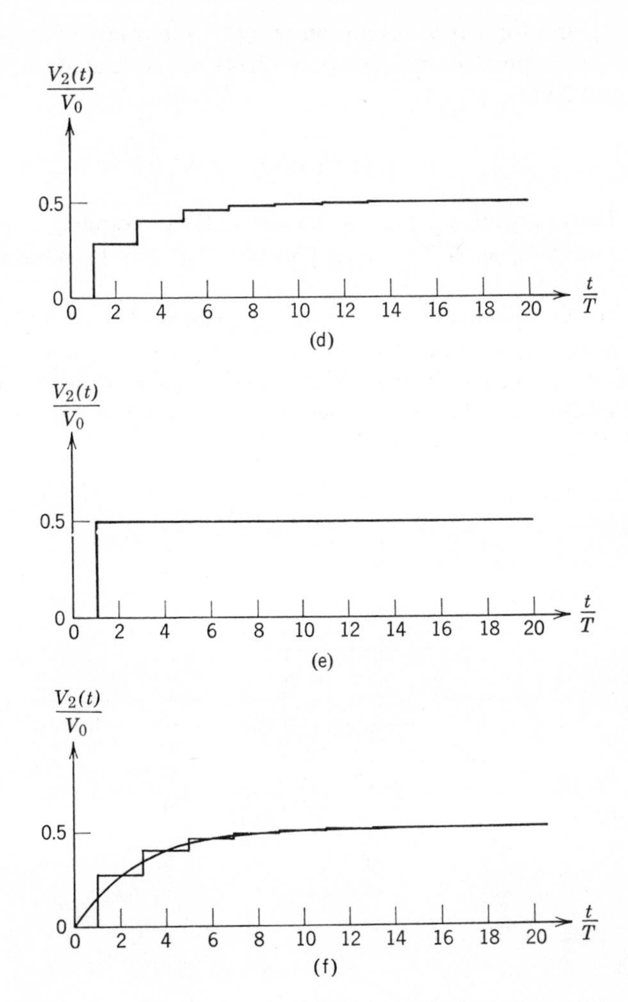

Figure 2.9 Transient response of the circuit of Fig. 2.8 for a step-function source and for various values of R_g and R_L:

(a) $R_g = 0.2R_0$, $R_L = \infty$
(b) $R_g = R_0$, $R_L = \infty$
(c) $R_g = 5R_0$, $R_L = \infty$
(d) $R_g = R_L = 0.2R_0$
(e) $R_g = R_L = R_0$
(f) $R_g = R_L = 5R_0$

$R_L \gg R_0$, and for a time resolution of $\gg 2T$, the transient response can be approximated by Equation (2.74) as illustrated in Figs. 2.9(c) and 2.9(f).

EXERCISES

1. Derive Equation (2.5) from Equation (2.1) by integrating by parts.

2. Derive Equation (2.9) using Equation (2.1) and the substitution $\xi \equiv t/T$.

3. Derive Equation (2.14) by utilizing Equations (2.1) and (2.11).

4. In the circuit of Fig. 2.1, find $V_C(t)$, if $V_g(t)$ is a single pulse with a height of 1 V and a width of T_1, by decomposing the pulse into a sum of a step function and a delayed step function as shown in Fig. 2.10. Sketch $V_C(t)$ for $T = RC/2$, RC, and $2RC$.

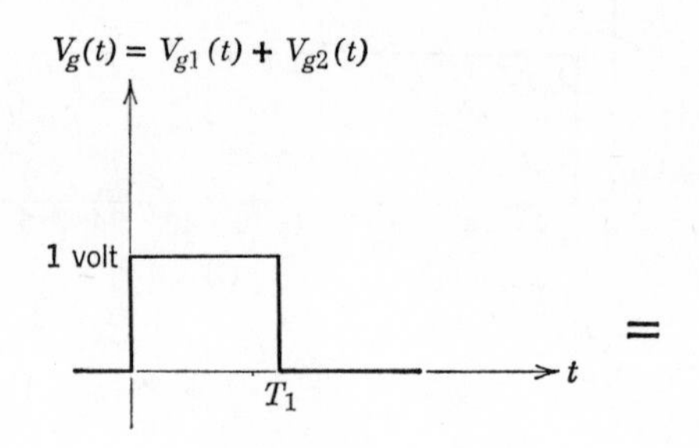

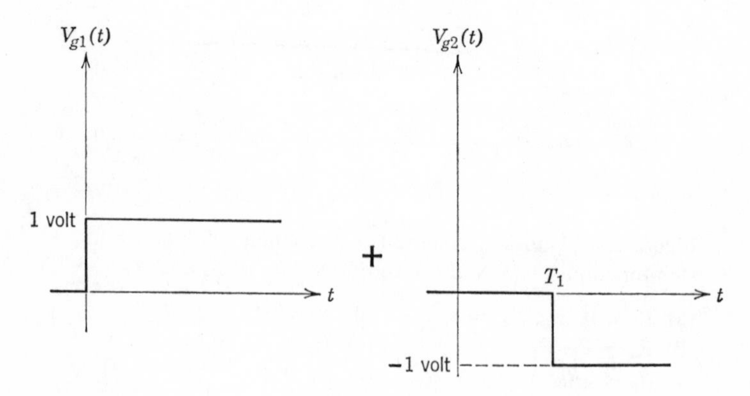

Figure 2.10

5. Find $V_L(t)$ and $V_R(t)$ in the circuit of Fig. 2.11 with $I(t < 0) = 0$.

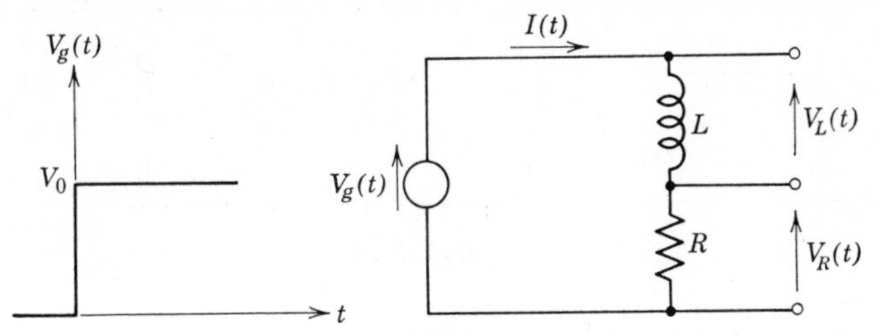

Figure 2.11

6. Find $I_C(t)$ and $I_R(t)$ in the circuit of Fig. 2.12 with $I_R(t < 0) = 0$.

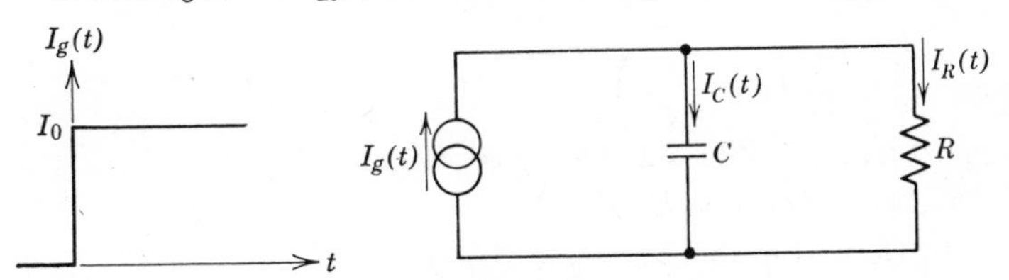

Figure 2.12

7. Find $I_L(t)$ and $I_R(t)$ in the circuit of Fig. 2.13 with $I_L(t < 0) = 0$.

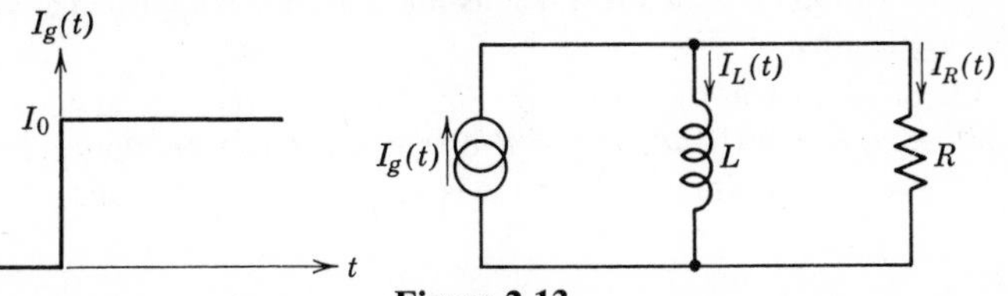

Figure 2.13

8. Show that in the circuit of Fig. 2.14 with $I_L(t < 0) = 0$ and $I_R(t < 0) = 0$, the voltage $V(t)$ can be written

(a) For $\dfrac{L}{R^2C} > 4$ as

$$V(t) = \frac{I_0}{CD} e^{-t/2RC} \sinh Dt,$$

where $D^2 \equiv \left(\dfrac{1}{2RC}\right)^2 - \dfrac{1}{LC}$;

(b) For $\dfrac{L}{R^2C} = 4$ as

$$V(t) = \frac{I_0}{C} t\, e^{-t/2RC};$$

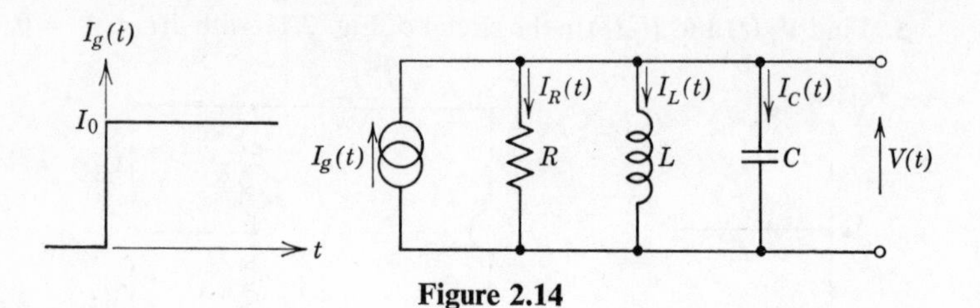

Figure 2.14

(c) For $\dfrac{L}{R^2C} < 4$ as

$$V(t) = \frac{I_0}{\omega_0 C} e^{-t/2RC} \sin \omega_0 t,$$

where $\omega_0{}^2 \equiv \dfrac{1}{LC}\left(1 - \dfrac{L}{4R^2C}\right)$.

Sketch $V(t)$ for $L/R^2C = 2$ utilizing the results of Fig. 2.5.

9. Derive Equation (2.51).

10. Derive Equations (2.52), (2.53), and (2.54).

11. Sketch $V_2(t)$ in the circuit of Fig. 2.8, if $R_g = 0$ and $R_L = \infty$.

12. Show that in the circuit of Fig. 2.8, if $R_g \ll R_0$ and $R_L \ll R_0$, for a time resolution of $\gg 2T$ the transmission line can be approximated by a series inductance of TR_0.

13. Show that in the circuit of Fig. 2.15 with zero initial conditions, the impedance $\mathscr{L}\{V(t)\}/\mathscr{L}\{I(t)\}$ is R if the box Z is a resistance R; it is $1/Cs$ if Z is a capacitance C; and it is Ls if Z is an inductance L.

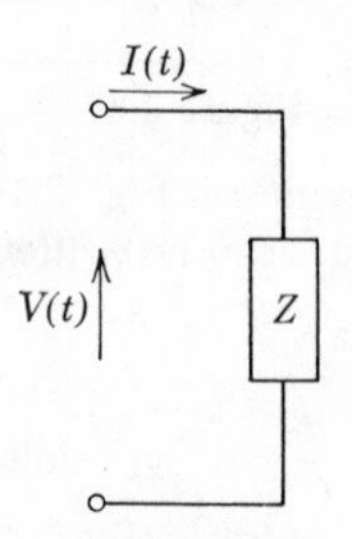

Figure 2.15

14. Use the results of Exercise 13 and the Laplace transform of the unit step to arrive at Equation (2.38).

ooooo

3

ooooo

Transient Response of Cascaded Circuits and Ladder Networks

ooooooooooooooooooooooooooooooooo

THE TRANSFER FUNCTION

In circuits consisting of resistors, capacitors, and inductors, the Laplace transform of the voltage or of the current output for a step-function or a δ-function input can always be written as a ratio of two polynomials. In the case when the initial conditions are zero, this ratio is a characteristic of the particular network. If, in addition to zero initial conditions, the source $V_g(t)$ or $I_g(t)$ is a unit δ-function, $\delta(t)$, the ratio is designated as the transfer function $H(s)$, and its inverse Laplace transform $\mathscr{L}^{-1}\{H(s)\} = h(t)$ as the impulse response, or δ-function response. In general, such a transfer function can always be written in the form

$$H(s) = K_1 \frac{(s + b_1)(s + b_2) \cdots (s + b_n)}{(s + a_1)(s + a_2) \cdots (s + a_m)}, \tag{3.1}$$

or in the form

$$H(s) = \frac{B_0 + B_1 s + B_2 s^2 + \cdots + B_n s^n}{A_0 + A_1 s + A_2 s^2 + \cdots + A_m s^m}, \tag{3.2}$$

where $K_1, b_1, \ldots, b_n, a_1, a_2, \ldots, a_m, B_0, B_1, \ldots, B_n, A_0, A_1, \ldots, A_m$ are constants.

THE ELMORE DELAY AND THE ELMORE RISETIME

Often the magnitudes of the a's and b's in Equation (3.1) fall into two widely separated groups. In this case, it is convenient to separate $H(s)$ into two factors:

$$H(s) = H_1(s)H_2(s), \tag{3.3}$$

where $H_1(s)$ contains all small a's and b's, $H_2(s)$ contains all the large ones, and constant factors are distributed such that

$$\lim_{s \to 0} H_2(s) = 1. \tag{3.4}$$

If this separation is performed, $H_1(s)$ describes the network for long times, $H_2(s)$ for short times. In the following, attention will be focused on short times and on $H_2(s)$. Reference will be also made to the corresponding impulse response

$$h_2(t) \equiv \mathscr{L}^{-1}\{H_2(s)\}, \tag{3.5}$$

and to the step-function response $e_2(t)$, i.e., to the response for an input of a unit step-function $u(t)$. Since the initial conditions are zero,

$$e_2(t) = \int_0^t h_2(t')\,dt' = \mathscr{L}^{-1}\left\{\frac{H_2(s)}{s}\right\}. \tag{3.6}$$

As a result of the final value theorem,

$$\lim_{t \to \infty} e_2(t) = 1, \tag{3.7}$$

thus, for a unit step-function input the output of a network described by $H_2(s)$ will be zero for $t < 0$ and change to 1 for large times. This transition can be characterized by a 10% to 90% risetime, which is the time between reaching the 10% and the 90% points of the final unit height. This time is easy to measure; it is, however, difficult to compute for all but the simplest networks.

If the impulse response $h_2(t)$ is known, then in many cases a better approach is to define a delay T_D and arise time T_R of the step-function response in terms of the first and second moments of the impulse response (see Elmore and Sands in references):

$$T_D \equiv \int_0^\infty t\, h_2(t)\, dt \tag{3.8}$$

and

$$T_R^2 \equiv 2\pi \int_0^\infty (t - T_D)^2 h_2(t)\, dt \tag{3.9}$$

that can also be written as

$$T_R^2 = 2\pi\left[\int_0^\infty t^2 h_2(t)\, dt - T_D^2\right] \tag{3.10}$$

by utilizing

$$\int_0^\infty h_2(t)\, dt = 1 \tag{3.11}$$

which follows from Equations (3.4) and (3.5). The delay T_D known as the Elmore delay and the risetime T_R known as the Elmore risetime are particularly useful if the step-function response $e_2(t)$ is monotonic, i.e., if $h_2(t) \geq 0$ for all t. If $e_2(t)$ is not monotonic, $h_2(t)$ is negative at some t; hence T_R can become zero, or even negative, and thus lose its utility. For monotonic $e_2(t)$, however, the Elmore risetime and the 10% to 90% risetime are, as a rule, reasonably close.

Several $H_2(s)$, $h_2(t)$, $e_2(t)$, T_D, T_R, and 10% to 90% risetimes are tabulated in Table 3.1; $h_2(t)$ and $e_2(t)$ are also shown in Fig. 3.1.

An inspection of Fig. 3.1 shows that as n increases, the impulse response of $H_2(s) = a^n/(s + a)^n$ approaches a gaussian. It can be shown, although it will not be proven here, that, in analogy to the central limit theorem of statistics, the impulse response of a network whose $H_2(s)$ is a product of n factors becomes a gaussian as $n \to \infty$ (see Elmore in references).

CASCADED CIRCUITS

If a signal passes through several cascaded circuits with transfer functions that are independent of each other, the resulting transfer

Table 3.1 Transfer functions, impulse responses, step-function responses, Elmore-delays, Elmore-risetimes, and 10% to 90% risetimes of selected circuits.

Circuit	Transfer function $H_2(s)$	Impulse response $h_2(t)$	Step-function response $e_2(t)$	Elmore Delay T_D	Elmore Risetime T_R	10% to 90% Risetime
n-stage	$\left(\frac{a}{s+a}\right)^n$	$\frac{a^n t^{n-1} e^{-at}}{(n-1)!}$	$1 - e^{-at}\left[1 + \frac{at}{1!} + \cdots + \frac{(at)^{n-1}}{(n-1)!}\right]$	$\frac{n}{a}$	$\frac{2.5\sqrt{n}}{a}$	See below for various values of n
1-stage	$\frac{a}{s+a}$	ae^{-at}	$1 - e^{-at}$	$\frac{1}{a}$	$\frac{2.5}{a}$	$\frac{2.2}{a}$
2-stage	$\left(\frac{a}{s+a}\right)^2$	$a^2 t e^{-at}$	$1 - e^{-at}(1 + at)$	$\frac{2}{a}$	$\frac{3.5}{a}$	$\frac{3.5}{a}$
3-stage	$\left(\frac{a}{s+a}\right)^3$	$\frac{a^3 t^2 e^{-at}}{2}$	$1 - e^{-at}\left(1 + at + \frac{a^2t^2}{2}\right)$	$\frac{3}{a}$	$\frac{4.3}{a}$	$\frac{4.3}{a}$
Critically damped	$\frac{a}{2}\frac{s+2a}{(s+a)^2}$	$\frac{a(1+at)e^{-at}}{2}$	$1 - e^{-at}\left(1 + \frac{at}{2}\right)$	$\frac{3}{2a}$	$\frac{3.3}{a}$	$\frac{3.2}{a}$
Gaussian	$e^{-sD}e^{(s/2a)^2}$ $e^{a^2D^2} \gg 1$	$\frac{a}{\sqrt{\pi}} e^{-a^2(t-D)^2}$	erf $[a(t - D)]$	D	$\frac{1.8}{a}$	$\frac{1.8}{a}$

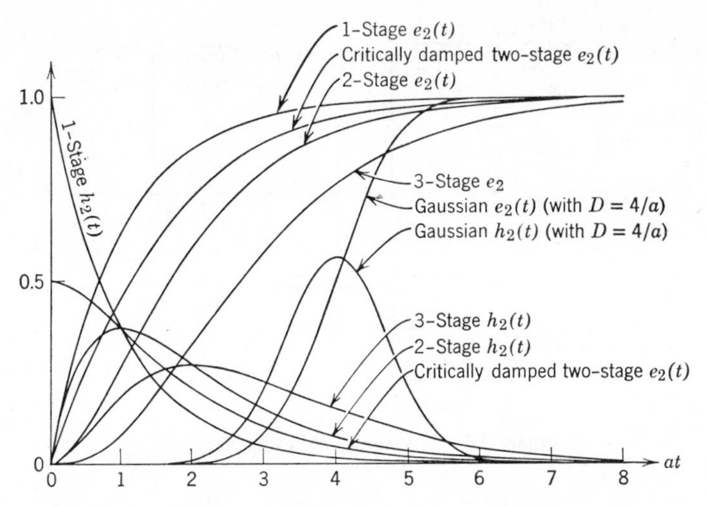

Figure 3.1 Impulse response $h_2(t)$ and step-function response $e_2(t)$ for the circuits of Table 3.1.

function is the product of the individual transfer functions. In this case, it can be shown that the resulting T_D is the sum of the individual T_D's and the resulting T_R^2 is the sum of the individual T_R^2's, i.e.,

$$T_{D_{\text{res}}} = \sum T_D \tag{3.12}$$

and

$$T_{R_{\text{res}}}^2 = \sum T_R^2 \tag{3.13a}$$

that can also be written as

$$T_{R_{\text{res}}} = \sqrt{\sum T_R^2}. \tag{3.13b}$$

The relationship (3.13) is strictly true if T_R is the Elmore risetime of Equation (3.9); it can be used, however, to approximate 10% to 90% risetimes with reasonable accuracy if the $e_2(t)$'s are monotonic, or are nearly so.

LADDER NETWORKS

Ladder networks of the type shown in Fig. 3.2 frequently arise in high-speed pulse circuits. In order to compute the transient for a

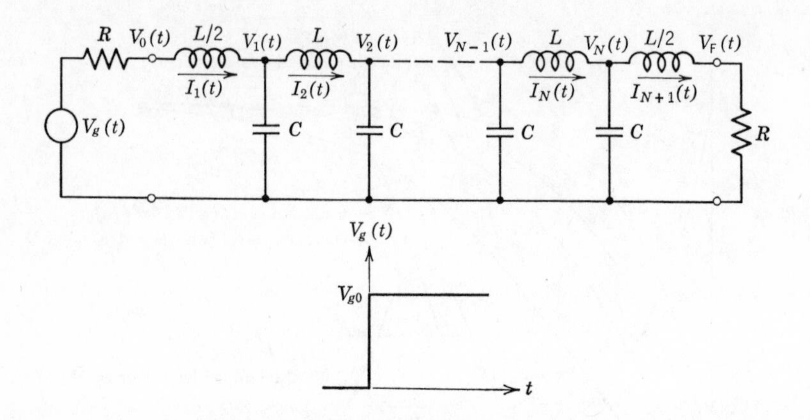

Figure 3.2 *N*-stage *L-C* ladder network.

step-function input, the following equations can be written:

$$V_0(t) = V_g(t) - RI_1(t) \tag{3.14}$$

$$I_1(t) = \frac{2}{L}\int [V_0(t) - V_1(t)]\, dt \tag{3.15}$$

$$V_j(t) = \frac{1}{C}\int [I_j(t) - I_{j+1}(t)]\, dt \qquad 1 \le j \le N-1 \tag{3.16}$$

$$I_{j+1}(t) = \frac{1}{L}\int [V_j(t) - V_{j+1}(t)]\, dt \qquad 1 \le j \le N-1 \tag{3.17}$$

$$V_N(t) = \frac{1}{C}\int [I_N(t) - I_{N+1}(t)]\, dt \tag{3.18}$$

$$I_{N+1}(t) = \frac{2}{L}\int [V_N(t) - V_F(t)]\, dt \tag{3.19}$$

$$V_F(t) = RI_{N+1}(t). \tag{3.20}$$

Unfortunately, even the simplest case of $N = 1$, i.e., a circuit with one capacitor and two inductors, leads to an equation that is cubic

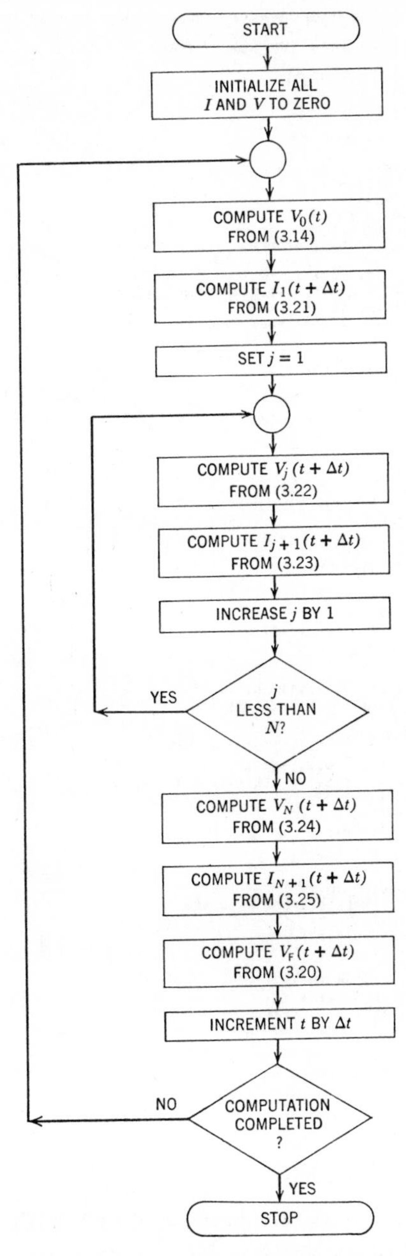

Figure 3.3 Flow-chart of the computer program for the transient analysis of the circuit shown in Fig. 3.2.

```
C     *TRANSIENT RESPONSE OF N-STAGE LADDER NETWORK TO STEP-FUNCTION INPUT*
C
C     *CALL PLOTTING ROUTINE AND MOVE PAPER*
      CALL STRTP1(10)
      CALL PLOT1(15.0,0.0,-3)
      REAL L,DELT
      REAL V(10),I(10)
      REAL VZERO,IZERO
      REAL VPLOT(10,802)
C
C     *READ AND WRITE L,DELT,N*
   99 READ(5,1)L,DELT,N
    1 FORMAT(2F10.5,I5)
      IF(L.EQ.0.0)GO TO 100
      WRITE(6,2)L,DELT,N
    2 FORMAT(' ','L=',F10.5,'DELT=',F10.5,'N=',I5)
C     *PLOT AXES*
      CALL PLOT1(0.0,0.0,+3)
      CALL AXIS1(0.0,0.0,'T',-1,8.0,0.0,0.0,1.0,10.0)
      CALL PLOT1(0.0,0.0,+3)
      CALL AXIS1(0.0,0.0,'V',+1,6.0,90.,0.0,0.20,4.0)
      CALL PLOT1(0.0,0.0,+3)
C
C     *INITIALIZE VARIABLES*
      INX=1
      VZERO=0.0
      IZERO=0.0
      DO 11 K=1,10
      V(K)=0.0
      I(K)=0.0
      DO 12 M=1,802
      VPLOT(K,M)=0
   12 CONTINUE
   11 CONTINUE
      NP1=N+1
C
C     *BEGIN COMPUTATION*
   21 CONTINUE
      T=INX*DELT
      VZERO=1.0-I(1)
      I(1)=I(1)+2.0*(VZERO-V(1))*DELT/L
      J=1
C
   22 CONTINUE
      V(J)=V(J)+(I(J)-I(J+1))*DELT
      I(J+1)=I(J+1)+(V(J)-V(J+1))*DELT/L
      J=J+1
      IF(J.LT.N)GO TO 22
C
      V(N)=V(N)+(I(N)-I(N+1))*DELT
      I(N+1)=I(N+1)+2.0*(V(N)-V(N+1))*DELT/L
      V(N+1)=I(N+1)
      DO 23 J1=1,NP1
      VPLOT(J1,INX)=V(J1)
   23 CONTINUE
C
      INX=INX+1
      IF(INX.LE.800)GO TO 21
C     *COMPUTATION COMPLETED,PLOT RESULTS*
      DO 33 J2=1,NP1
      CALL PLOT1(0.0,0.0,+3)
      DO 34 IX=1,800
      XPLOT=DELT*IX
      IF(XPLOT.GT.8.0)GO TO 35
      YPLOT=10.*VPLOT(J2,IX)
      CALL PLOT1(XPLOT,YPLOT,+2)
   34 CONTINUE
   35 CONTINUE
   33 CONTINUE
      CALL PLOT1(15.,0.0,-3)
C     *PLOTTING COMPLETED*
      GO TO 99
  100 CONTINUE
      CALL PLOT1(15.0,0.0,-3)
      CALL ENDP1
      STOP
      END
```

Figure 3.4 Fortran-IV computer program for the transient analysis of the circuit shown in Fig. 3.2.

in s. To avoid complex computations, a different method will be followed: The integrals will be converted to sums, and solution by a digital computer will be performed.

Equation (3.15) can be approximated as

$$I_1(t) \cong \frac{2}{L} \sum [V_0(t) - V_1(t)]\, \Delta t,$$

that can also be written as

$$I_1(t + \Delta t) = I_1(t) + \frac{2}{L} [V_0(t) - V_1(t)]\, \Delta t. \tag{3.21}$$

Similarly,

$$V_j(t + \Delta t) = V_j(t) + \frac{1}{C}\Big[I_j(t) - I_{j+1}(t)\Big]\, \Delta t \tag{3.22}$$

$$1 \le j \le N - 1$$

$$I_{j+1}(t + \Delta t) = I_{j+1}(t) + \frac{1}{L}\Big[V_j(t) - V_{j+1}(t)\Big]\, \Delta t \tag{3.23}$$

$$V_N(t + \Delta t) = V_N(t) + \frac{1}{C}\Big[I_N(t) - I_{N+1}(t)\Big]\, \Delta t \tag{3.24}$$

$$I_{N+1}(t + \Delta t) = I_{N+1}(t) + \frac{2}{L}\Big[V_N(t) - V_F(t)\Big]\, \Delta t. \tag{3.25}$$

Starting with zero initial conditions, these equations are solved by a program with the flow-chart shown in Fig. 3.3 and with the Fortran-IV program of Fig. 3.4. The resulting voltages are plotted for $1 \le N \le 5$ in Fig. 3.5(a), (b), and (c) for $L/R^2C = \frac{1}{6}, \frac{1}{3}$, and $\frac{2}{3}$, respectively. Voltages along the network for the case of $N = 5$ and $L/R^2C = \frac{2}{3}$ are shown in Fig. 3.5(d). The overshoot and ringing become more pronounced for increasing inductance L; it can be shown, however, that these effects are reduced for input signals with finite risetimes.

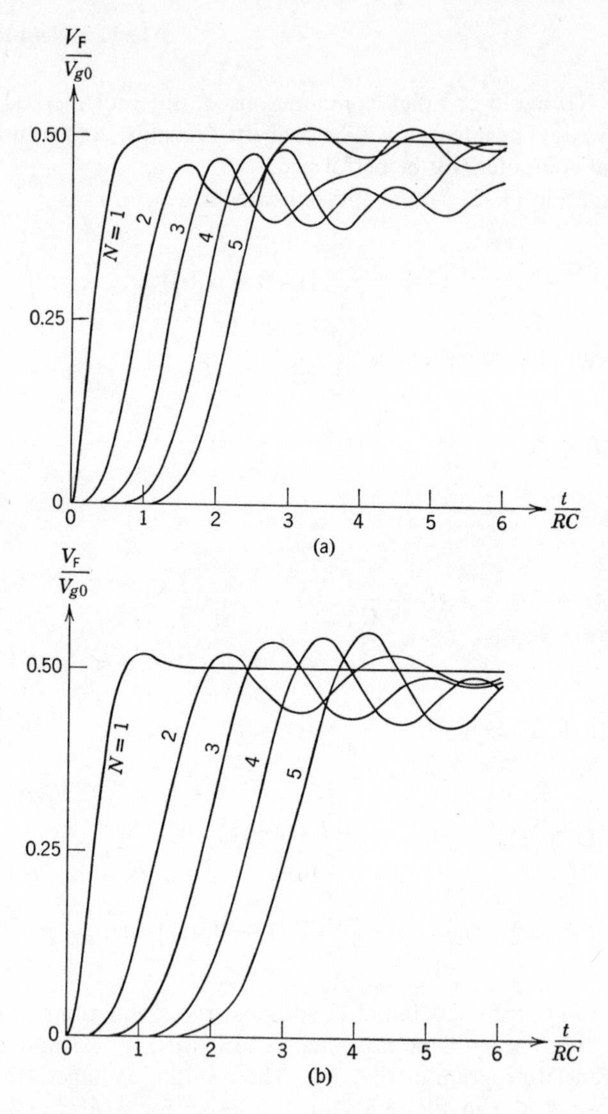

Figure 3.5 Transient response of the circuit of Fig. 3.2 for a unit step-function input. (a) Output voltage V_F/V_{g0} for $1 \leq N \leq 5$ and $L/R^2C = \frac{1}{6}$, (b) Output voltage V_F for $1 \leq N \leq 5$ and $L/R^2C = \frac{1}{3}$, (c) Output voltage V_F for $1 \leq N \leq 5$ and $L/R^2C = \frac{2}{3}$, (d) Voltages V_1 through V_F for $N = 5$ and $L/R^2C = \frac{2}{3}$.

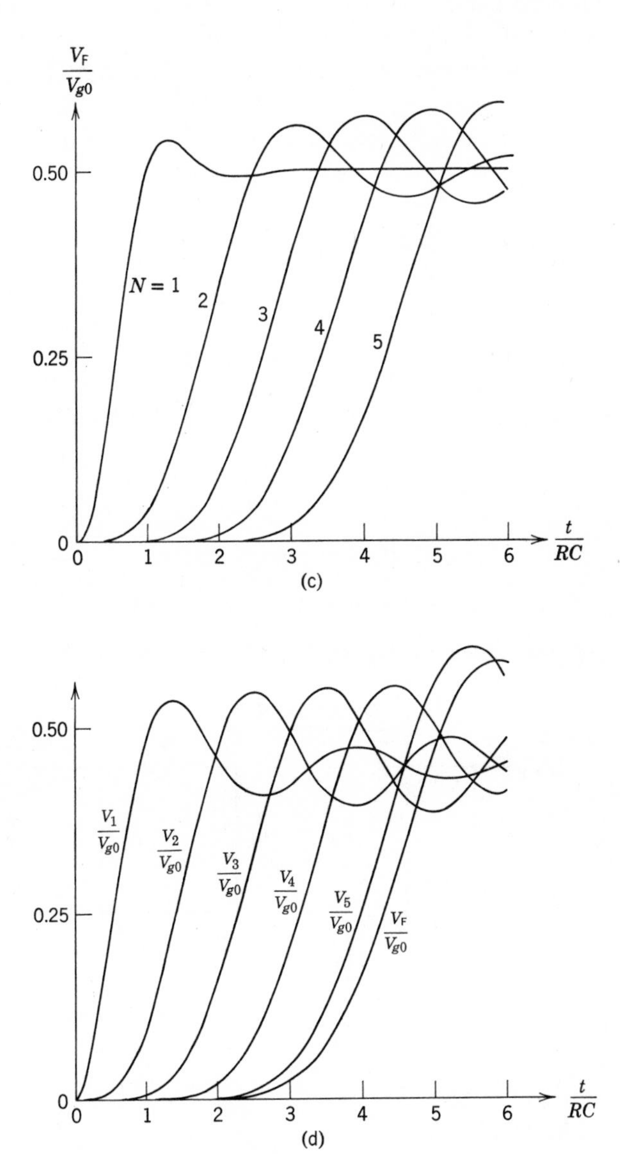

Figure 3.5 (Continued).

EXERCISES

1. Write the transfer function $\mathscr{L}\{V_L(t)\}/\mathscr{L}\{V_g(t)\}$ for the circuit of Fig. 2.4 with zero initial conditions.

2. Show that the transfer function $\mathscr{L}\{V_2(t)\}/\mathscr{L}\{I_g(t)\}$ of Figure 3.6

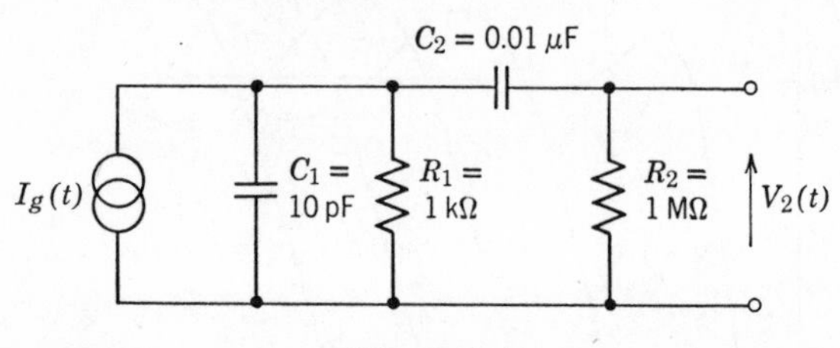

Figure 3.6

can be expressed with an error of less than 1% as $H_1(s)H_2(s)$, where

$$H_1(s) = \frac{R_1 s}{s + a_1},$$

$$H_2(s) = \frac{a_2}{s + a_2},$$

$$a_1 = 10^2/\text{second, and } a_2 = 10^8/\text{second.}$$

3. For the circuit of Figure 3.7, calculate the transfer function

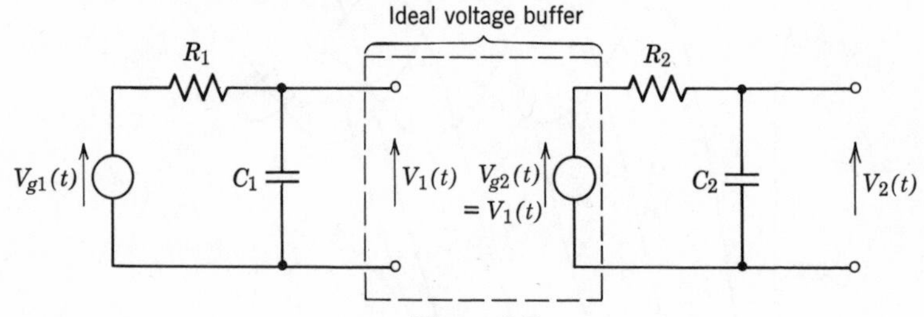

Figure 3.7

$\mathscr{L}\{V_2(t)\}/\mathscr{L}\{V_{g1}(t)\}$ and the corresponding impulse response for three cases:

(a) $R_1 = R_2,\ C_1 = C_2$
(b) $R_1 = 2R_2,\ C_1 = 2C_2$
(c) $R_1 = 2R_2,\ C_1 = C_2/2$.

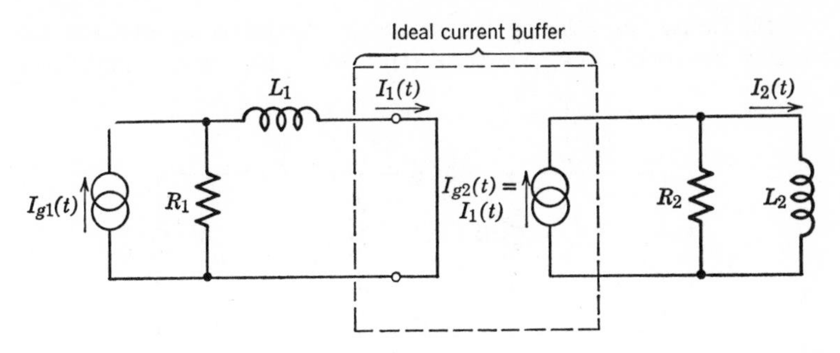

Figure 3.8

4. For the circuit of Figure 3.8, calculate the transfer function $\mathscr{L}\{I_2(t)\}/\mathscr{L}\{I_{g1}(t)\}$ and the corresponding impulse response for three cases:
(a) $R_1 = R_2$, $L_1 = L_2$
(b) $R_1 = 2R_2$, $L_1 = L_2/2$
(c) $R_1 = 2R_2$, $L_1 = 2L_2$.

5. Derive Equation (3.10) from Equations (3.4), (3.5), (3.8), and (3.9).

6. Show that for the entries of Table 3.1, $\int_0^\infty h_2(t)\,dt = 1$.

7. Derive the values of T_D and T_R of Table 3.1.

8. Demonstrate that Equation (3.13) is valid for

$$H_2(s) = \frac{a_1}{s + a_1}\,\frac{a_2}{s + a_2}$$

by breaking up $H_2(s)$ into two factors as shown. Assume $a_1 \neq a_2$, $\mathrm{Im}(a_1) = \mathrm{Im}(a_2) = 0$, $\mathrm{Re}(a_1) > 0$, $\mathrm{Re}(a_2) > 0$.

9. Derive the transfer function $\mathscr{L}\{V_{\text{out}}(t)\}/\mathscr{L}\{V_g(t)\}$ and the input impedance $Z_{\text{in}} \equiv \mathscr{L}\{V_g(t)\}/\mathscr{L}\{I_{\text{in}}(t)\}$ for the shaper circuit of Fig. 3.9 with $L/R = RC$. What is $V_{\text{out}}(t)$ for a $V_g(t) = \Phi_0\delta(t)$?

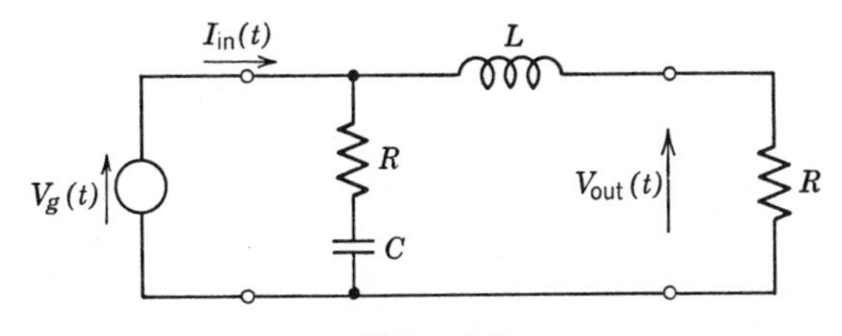

Figure 3.9

10. Derive, by utilizing the results of the preceding exercise, the transfer function $\mathscr{L}\{V_{\text{out}}(t)\}/\mathscr{L}\{V_g(t)\}$ and the input impedance

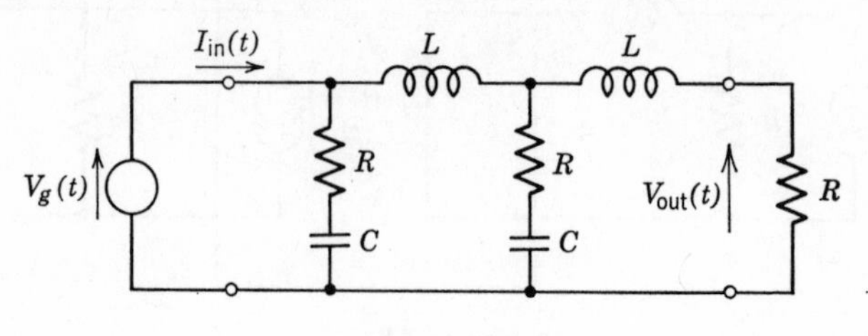

Figure 3.10

$Z_{\text{in}} \equiv \mathscr{L}\{V_g(t)\}/\mathscr{L}\{I_{\text{in}}(t)\}$ for the shaper circuit of Fig. 3.10 with $L/R = RC$. What is $V_{\text{out}}(t)$ for a $V_g(t) = \Phi_0\delta(t)$?

4

Real Components

Characteristics of ideal components were presented in Chapter 1. In reality, none of these exist individually and often several of them are unavoidably combined in one real component. In addition to the basic "desired" component (say, resistance), there are additional "parasitic" elements (such as inductance, capacitance) present. These parasitic elements can result in a significant change in the performance of the component in a circuit; hence, it is important to be aware of their existence and typical magnitudes.

RESISTORS

The resistance R, in ohms, of a slab of material of length L and cross-sectional area WH (Fig. 4.1) is

$$R = \rho \frac{L}{WH}, \tag{4.1}$$

where ρ is the resistivity in ohm-meter, usually a function of temperature. For copper at room temperature, $\rho \approx 1.7 \times 10^{-8}$ ohm-meter.

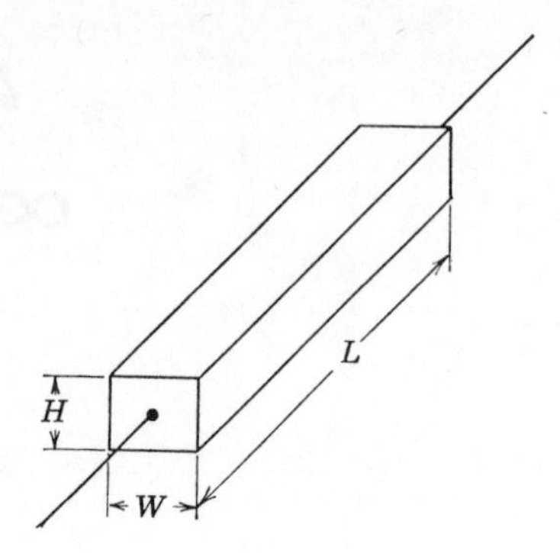

Figure 4.1 A slab-resistor.

Resistors made of carbon composition are the most common in pulse circuits. The desired resistance is attained by varying the composition (resistivity) of the material. These resistors are inexpensive, accurate to a few percent, low in inductance, and are available with maximum power dissipations of up to 2 watts. Higher precisions and greater power dissipations are available in wire-wound resistors where the desired resistance is achieved by appropriate choice of the material and of the length of the wire wound in a spiral on a ceramic body: a structure which possesses undesirable inductance; the inductance is reduced ("non-inductive wirewound resistors") if the winding is in a double spiral. With suitable construction, wirewound resistors can achieve better than 0.001% precision at a constant temperature. Moderately good accuracy ($\approx$1%) and low inductance are achieved in metal-film resistors: these consist of a spiral of metal film deposited on the surface of a ceramic cylinder; power dissipations of up to 2 watts are

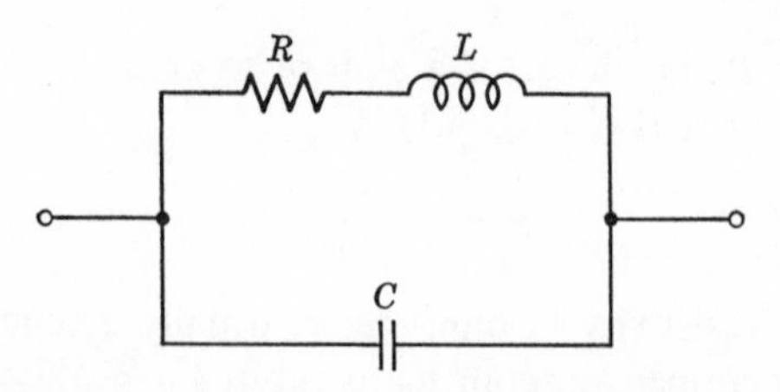

Figure 4.2 Equivalent circuit of a real resistor.

available. In addition to these, many additional types are available, such as those designed for high voltage, high temperature, etc.

A real resistor, in addition to its resistance, has parasitic elements, primarily inductance and capacitance (Fig. 4.2); typical values of these are shown in Table 4.1.

Table 4.1 Properties of real resistors.

Type	Typical R	Maximum Dissipation	Typical L	Typical C
Composition Carbon	2 Ω to 20 MΩ	$\frac{1}{8}$ W to 2 W	10 nH	0.3 pF
Wirewound	1 Ω to 100 kΩ	1 W to 500 W	1 μH to 100 μH	1 pF
Noninductive Wirewound	5 Ω to 10 kΩ	10 W to 200 W	0.1 μH to 1 μH	1 pF to 4 pF
Metal film	10 Ω to 2 MΩ	$\frac{1}{8}$ W to 2 W	50 nH	0.3 pF

CAPACITORS

The capacitance C, in farads, between two plates of size $W \times L$ each, separated by a distance D (Fig. 4.3), for $W \gg D$ and $L \gg D$ is

$$C = \epsilon_0 \epsilon_r \frac{WL}{D}. \tag{4.2}$$

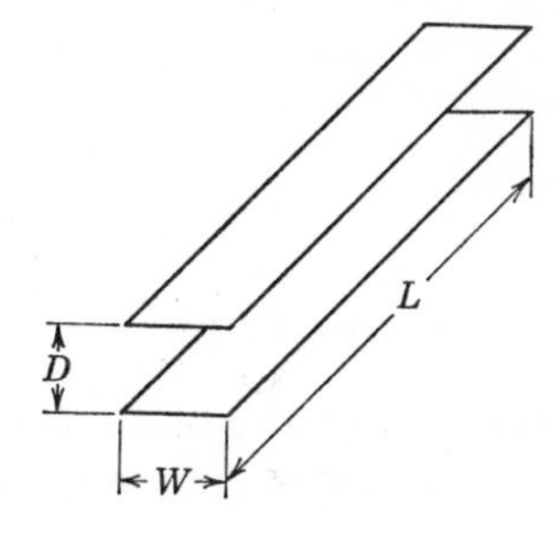

Figure 4.3 Parallel-plate capacitor.

Table 4.2 Properties of real capacitors.

Type	C	R_p	$\tau_s \equiv CR_s$	L
Air	0.1 pF to 500 pF	$>10^{14}\ \Omega$	<1 ns	1 nH to 10 nH
Mica	5 pF to 10,000 pF	$>10^{10}\ \Omega$	<1 ns	1 nH to 10 nH
Ceramic	5 pF to 2 μF	$>10^{9}\ \Omega$	<1 ns	1 nH to 10 nH
Paper	1000 pF to 20 μF	$>10^{8}\ \Omega$	$\approx 1\ \mu$s	10 nH

Here ϵ_0 is the permittivity of the vacuum: $\epsilon_0 = 8.85434 \times 10^{-12}$ F/m; ϵ_r is the relative permittivity, or dielectric constant, of the material between the plates. If $W \gg D$ does not hold, a better approximation is given by (see Morse and Feshbach in references)

$$C = \epsilon_0 \epsilon_r \frac{WL}{D}\left[1 + \frac{D}{\pi W} \ln\left(\frac{\pi e W}{D}\right)\right]. \tag{4.3}$$

Some of the more common dielectric materials are vacuum, air, mica, ceramic, and paper. Typical properties of various capacitors are shown in Table 4.2 for the equivalent circuit of Fig. 4.4.

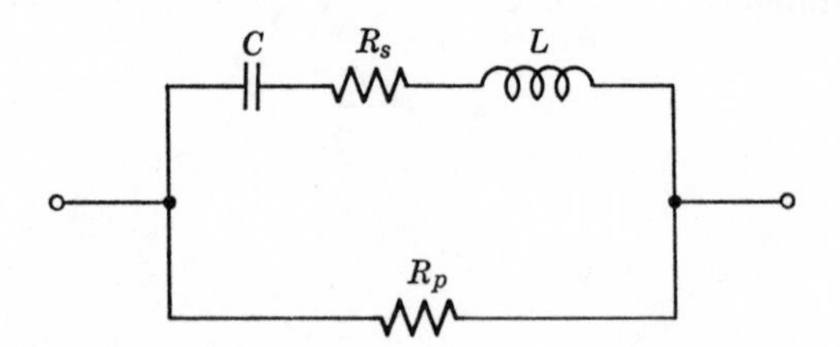

Figure 4.4 Equivalent circuit of a real capacitor.

INDUCTORS

The inductance of an inductor is a function of its shape, size, and core material. The inductance L, in henrys, of a single-layer coil of

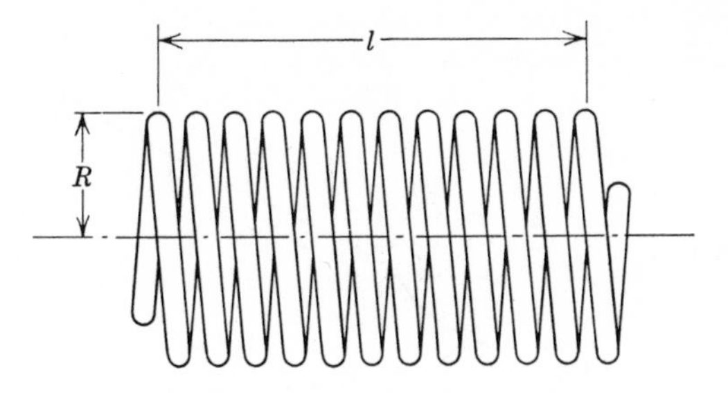

Figure 4.5 A single-layer solenoid.

length l, with radius $R \ll l$, and with N turns on a material with relative permeability of μ_r (Fig. 4.5) is

$$L \cong \mu_0 \mu_r \frac{N^2 \pi R^2}{l}, \tag{4.4}$$

where $\mu_0 = 4\pi \times 10^{-7}$ H/m is the permeability of vacuum.

An equivalent circuit for real inductors is shown in Fig. 4.6; values of inductance L, of $\tau_0 \equiv \sqrt{LC}$, and of $Q_L \equiv \tau_0/RC$ are shown in Fig. 4.7 for tubular inductors of $\sim\frac{1}{2}$-inch length.* Many magnetic materials are non-linear; hence, L can be a function of the current.

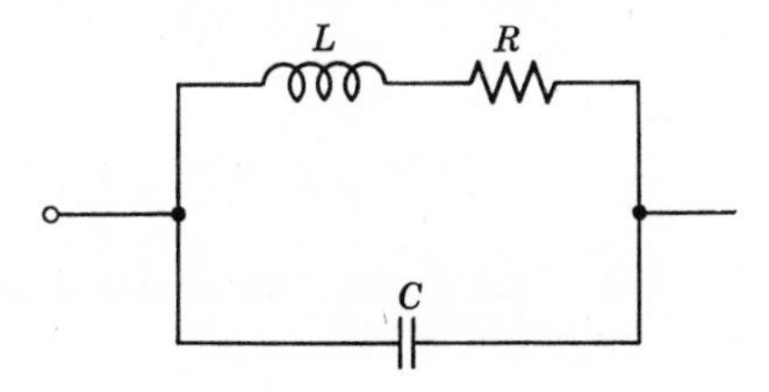

Figure 4.6 Equivalent circuit of a real inductor.

TRANSFORMERS

A real transformer has parasitic inductances, capacitances, and resistances as shown in a simple equivalent circuit in Fig. 4.8. Here all parasitic components are shown on the right side; they can be

* R. W. Miller Co. Molded R. F. chokes.

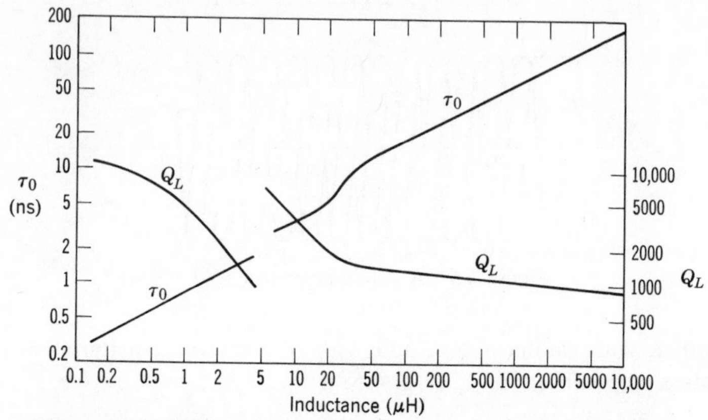

Figure 4.7 Values of τ_0 and Q_L for ~$\frac{1}{2}$-inch-long tubular inductors with air-core (below 5 μH), and with iron-core (above 5 μH).

transformed to the left side, if so desired, by appropriate scaling by N^2 (see Exercises 10 and 11 in Chapter 1).

Shunt inductance L_p is the parallel combination of the inductance of the right-side winding and N^2 times that of the left-side winding. Resistors R_1 and R_2 represent the resistances of the windings and losses in the iron core; in high-speed circuits these can be usually neglected. Capacitor C represents stray capacitances; leakage inductances L_1 and L_2 result from imperfect coupling between the two windings.

In pulse circuits, it is often possible to separate the "fast" and "slow" components, resulting in the equivalent circuits of Figs.

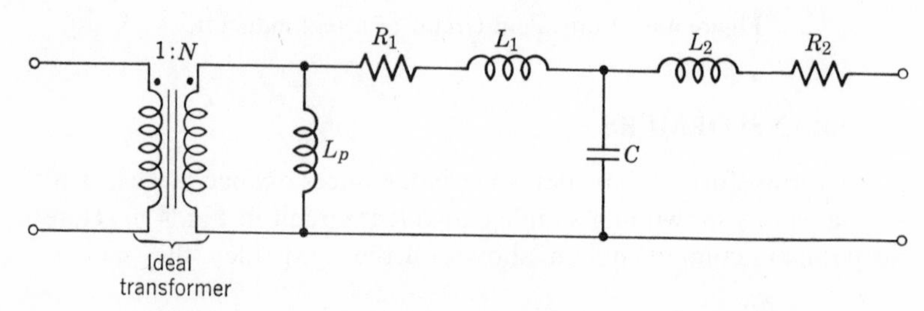

Figure 4.8 Equivalent circuit of a real transformer.

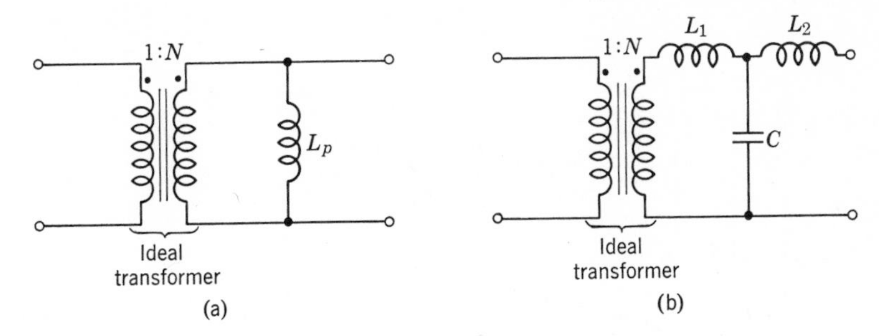

Figure 4.9 Simplification of the circuit of Fig. 4.8 for: (a) Long times. (b) Short times.

4.9(a) and (b) for long and short times, respectively. For good long-time performance it would be advantageous to have a large L_p, and for good short-time performance small L_1, L_2, and C. Large L_p, however, implies many turns—which increases L_1, L_2, and C. This dilemma is somewhat alleviated if the transformer winding is made of a coaxial transmission line which is then terminated by its characteristic impedance.

TRANSMISSION LINES

The characteristic impedance of a lossless transmission line with a dielectric of dielectric constant ϵ_r is $R_0 = T/C_\epsilon$, where T and C_ϵ are the time delay and capacitance, respectively. The velocity of signal propagation in a medium with a relative permeability of unity and a relative dielectric constant of ϵ_r is $c/\sqrt{\epsilon_r}$, where c is the velocity of light. Thus, the propagation time T for a length l is $T = l\sqrt{\epsilon_r}/c$, and the characteristic impedance $R_0 = l\sqrt{\epsilon_r}/cC_\epsilon$. The capacitance can be written as $C_\epsilon = \epsilon_r C_{\mathrm{VACUUM}}$; hence, the characteristic impedance

$$R_0 = \frac{l}{C_{\mathrm{VACUUM}}\, c\sqrt{\epsilon_r}} = \frac{(R_0)_{\mathrm{VACUUM}}}{\sqrt{\epsilon_r}}. \tag{4.5}$$

Characteristic impedances of practical vacuum-dielectric transmission lines consisting of two identical parallel plates, a plate and an infinite plane, two coaxial-cylinders, a wire and an infinite plane, and two parallel wires are shown in Fig. 4.10. If the dielectric is other than vacuum, R_0 should be divided by $\sqrt{\epsilon_r}$.

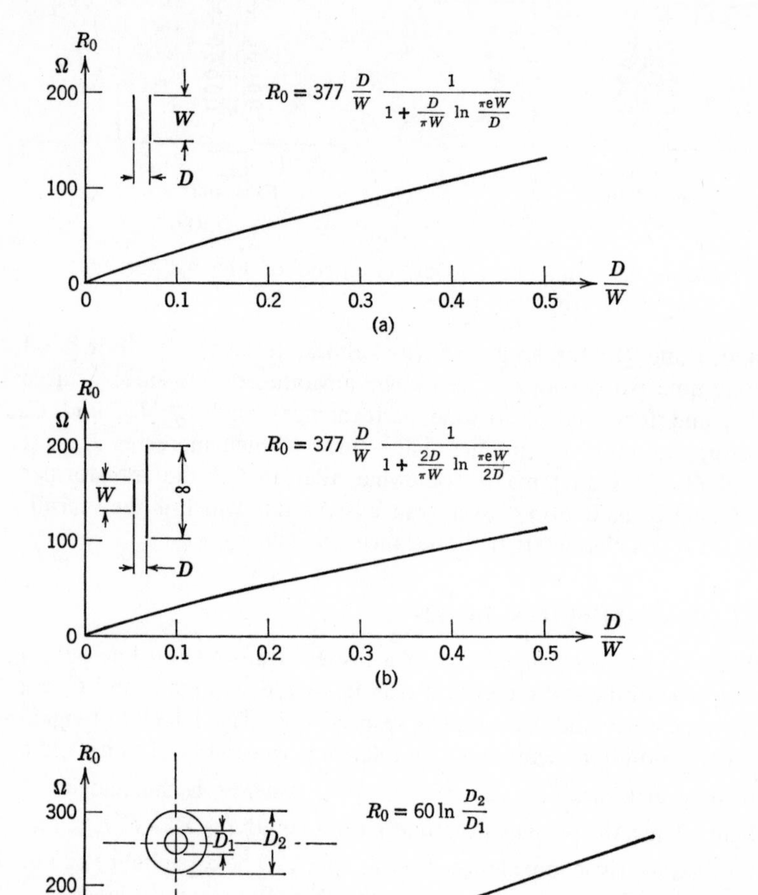

Figure 4.10 Characteristic impedance of transmission lines of various configurations with vacuum dielectric.

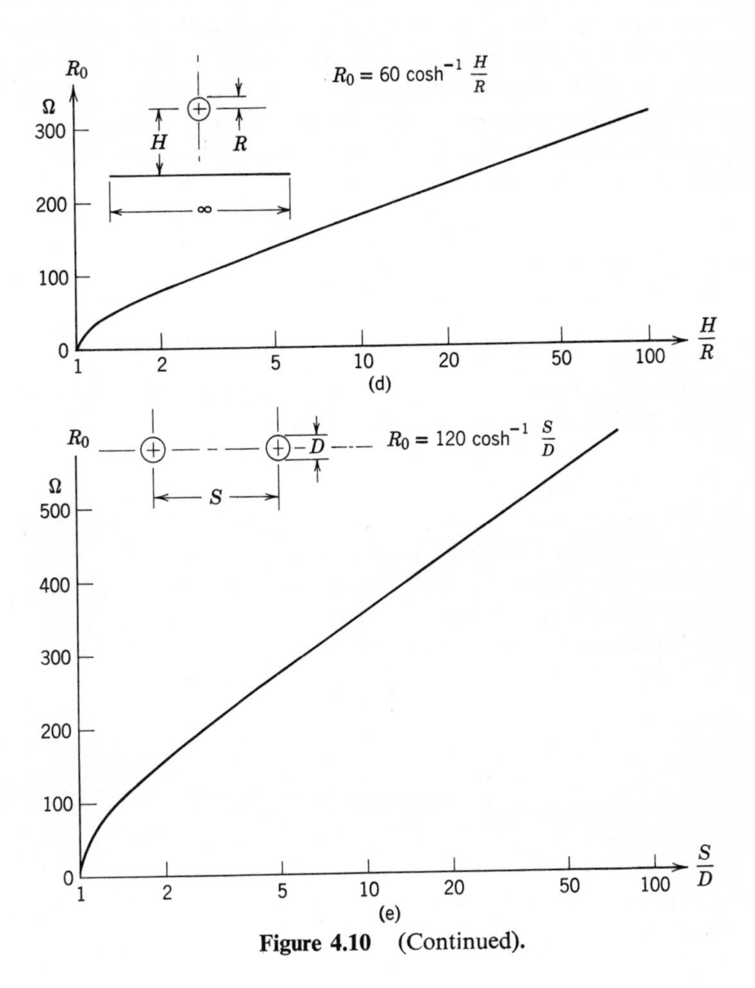

Figure 4.10 (Continued).

Real transmission lines also have losses as a result of resistance of the conductors, skin effect, dielectric losses, and resonance effects due to finite diameter. The most significant of these for risetimes slower than about a nanosecond is usually the skin effect. Due to skin effect only, the step-function response $e(t)$ of a coaxial transmission line with characteristic impedance R_0 and with an inner conductor of diameter D_1 and of specific resistivity ρ can be written as (see Wigington and Nahman in references)

$$e(T) = 1 - \mathrm{erf}(1/\sqrt{T}), \tag{4.6}$$

where

$$T \equiv \frac{16\pi^2 R_0^2 D_1^2}{\rho\mu l^2} t. \tag{4.7}$$

The function $e(T) = 1 - \mathrm{erf}(1/\sqrt{T})$, shown in Fig. 4.11, possesses a relatively fast initial rise which slows as $e(T)$ approaches its final value of unity; the 10% to 90% risetime is approximately 30 times the 10% to 50% risetime.

Unfortunately, due to other losses, transients based on Fig. 4.11 and on Equations (4.6) and (4.7) can be considered only crude approximations of those of an actual coaxial cable; the actual 10% to 90% risetime may be different by a factor of 5 (either way!) from that predicted by Equations (4.6) and (4.7).

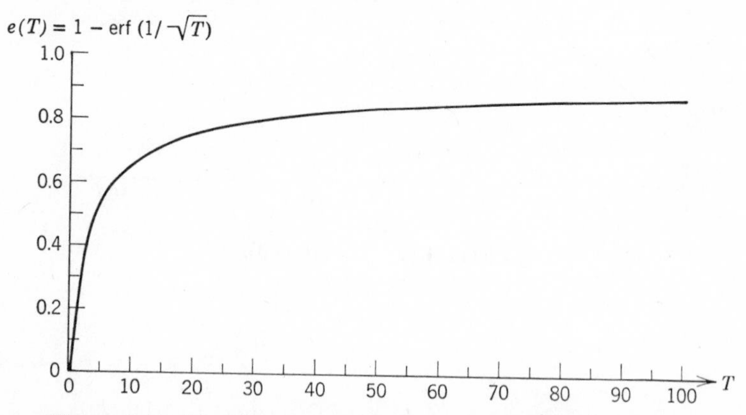

Figure 4.11 Transient response of a transmission line for a unit-step input, taking skin-losses into account.

EXERCISES

1. Use Equation (4.1) to calculate the resistance of a round copper wire with a diameter of 0.09 inch ($\approx$2.3 mm) and a length of 100 feet ($\approx$30 m).

2. Calculate the capacitance of a capacitor consisting of two plates of size 0.1 inch × 1 inch (2.56 mm × 25.6 mm) spaced at a distance of 0.05 inch (1.28 mm) in air ($\epsilon_r \approx 1$), using first Equation (4.2), then Equation (4.3).

3. Use the data of Fig. 4.7 to calculate the capacitance C and the resistance R of the inductor equivalent circuit (Fig. 4.6) for $L = 1\ \mu$H, 100 μH, and 10 mH.

4. Figure 4.12 shows a pulse transformer used for matching a 50-Ω source to a 200-Ω load. The transformer can be described by the equivalent circuit of Fig. 4.9 with $N = 2$, $L_p = 50\ \mu$H, $L_1 = L_2 = 0.1\ \mu$H, and $C = 15$ pF. What is the decay time constant of V_L? Utilize the data of Fig. 3.5 to compute the 10% to 90% risetime of V_L. Is the use of Fig. 4.9 justified?

5. Derive Equation (4.5) from Equation (4.2) by using the relationship $R_0 = T/C_\epsilon$.

6. Calculate the characteristic impedance of a coaxial cable with $D_1 = 0.09$ inch, $D_2 = 0.285$ inch, and $\epsilon_r = 2$.

7. Calculate the characteristic impedance of a coaxial cable with $D_1 = 0.003$ inch (#40 wire), $D_2 = 1$ inch, and $\epsilon_r = 1$.

8. What outer diameter D_2 would be needed to construct a coaxial cable with a characteristic impedance of (a) 500 Ω and (b) 1000 Ω, if $D_1 = 0.003$ inch (#40 wire) and $\epsilon_r = 1$?

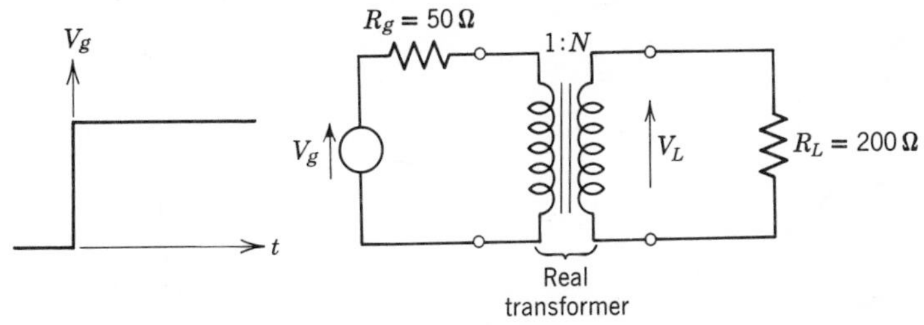

Figure 4.12

9. Show that for $S/D \gg 1$, the characteristic impedance of a transmission line of two parallel wires [Fig. 4.10(e)] can be approximated as

$$R_0 = \frac{120}{\sqrt{\epsilon_r}} \ln \frac{2S}{D}.$$

10. Use symmetry considerations to derive the results of Fig. 4.10(e) from those of Fig. 4.10(d), or vice versa.

11. Calculate the 10% to 50% and the 10% to 90% risetime of the coaxial cable of Exercise 6 above due to skin effect only. Assume a length of 100 feet ($\approx$30 m) and a copper center conductor ($\rho \approx 1.72 \times 10^{-8}$ ohm-meter); $\mu = \mu_0 = 4\pi \times 10^{-7}$ H/m.

5

Junction Diodes

Semiconductor devices have been finding increasing utilization in electronic circuits.* Semiconductor materials, such as germanium and silicon, have specific resistivities between that of conductors and that of insulators. For example, the specific resistivity of copper is $\rho = 1.7 \times 10^{-6}$ ohm-cm, that of glass is typically 10^{10} ohm-cm, while the specific resistivity of intrinsic (pure) germanium is $\rho \approx 47$ ohm-cm and that of intrinsic silicon is $\rho \approx 2.3 \times 10^{5}$ ohm-cm.

When an impurity from the fifth column of the periodic table, such as antimony, is added to an intrinsic semiconductor material, excess free electrons are produced, and an N-type semiconductor results. Similarly, if the added impurity is an element from the third column of the periodic table, such as gallium, a deficiency of free electrons, an excess of "holes", is created and a P-type material results.

CHARACTERISTICS

When an N-type and a P-type semiconducting material are joined, the resulting diode junction exhibits different electrical properties for opposite directions of current (Fig. 5.1). For an ideal diode with the

* In what follows here, only a brief summary of semiconductor characteristics is given. More details can be found in the references.

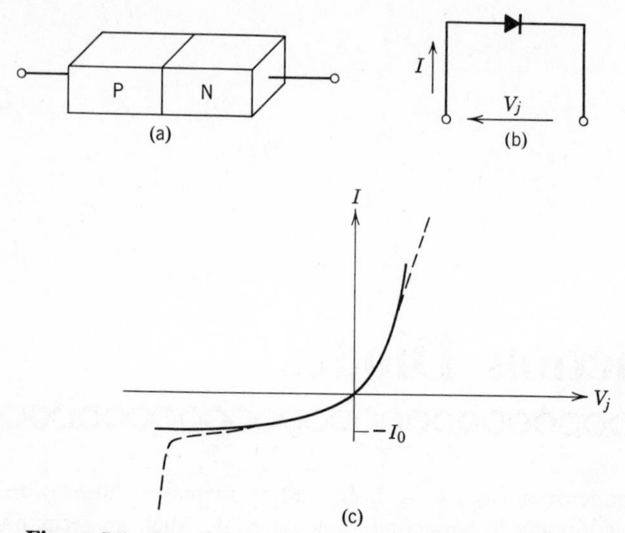

Figure 5.1 Junction diode. (a) Formation of the junction by joining a P-type and an N-type semiconductor. (b) Junction diode symbol. (c) Current versus voltage characteristics; Equation (5.1) is shown by solid line, characteristic of a real diode by broken line.

polarities as shown, the current through the junction is given by

$$I = I_0(e^{V_j/V_T} - 1), \tag{5.1}$$

where I is the current through the junction, V_j is the external voltage applied to the junction, and I_0 is the reverse saturation current (a constant for a given diode at a fixed temperature). Voltage V_T is

$$V_T = n\frac{kT}{q} \tag{5.2}$$

where $k = 1.38 \times 10^{-23}$ Ws/°K is the Boltzmann constant, T is the absolute temperature in °K, and $q = 1.6 \times 10^{-19}$ coulomb is the charge of the electron. The dimensionless multiplier n is typically 1 to 1.5 for germanium and 1.5 to 2 for silicon diodes. Thus, V_T is usually between 25 mV and 50 mV at room temperature.

Equation (5.1) indicates that for large negative V_j, the diode current approaches $-I_0$; the magnitude of I_0 is typically a few

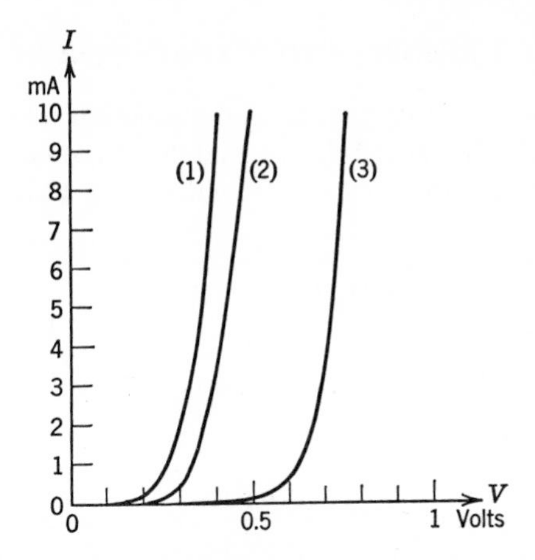

Figure 5.2 Forward characteristics of three junction diodes: (1) Germanium. (2) Metal-silicon. (3) Silicon.

nanoamperes for small silicon diodes. In reality, in addition to this current, a surface leakage current and a current resulting from junction breakdown at some negative voltage are also present [see Fig. 5.1(c)]. In the remainder of Chapter 5, however, these modifications will be neglected and Equation (5.1) will be utilized throughout. Measured forward characteristics of a germanium, a metal-silicon, and a silicon diode are shown in Fig. 5.2.

An important small-signal (incremental) dc characteristic of the diode, the incremental resistance r_i, is defined as

$$r_i \equiv \frac{dV_j}{dI}, \tag{5.3}$$

which, by using Equation (5.1), can be written as,

$$r_i \equiv \frac{dV_j}{dI} = \frac{\mathrm{V_T}}{I_0}\,\mathrm{e}^{-V_j/\mathrm{V_T}}, \tag{5.4a}$$

or as

$$r_i = \frac{\mathrm{V_T}}{I + I_0}. \tag{5.4b}$$

STORED CHARGE AND CAPACITANCES

In addition to the dc properties of the junction, stored charges are also important. The stored charge, and the concomitant capacitance, can be separated into a current dependent and a voltage dependent part:*

$$Q = Q_d(I) + Q_t(V_j) \tag{5.5}$$

and

$$C \equiv \frac{dQ}{dV_j} = C_d(I) + C_t(V_j), \tag{5.6}$$

where

$$C_d \equiv \frac{dQ_d}{dV_j} \tag{5.7a}$$

and

$$C_t \equiv \frac{dQ_t}{dV_j}. \tag{5.7b}$$

The charge Q_d and the capacitance C_d are due to the movement of carriers by diffusion. The charge Q_d, to a good approximation, is proportional to the junction current I:

$$Q_d = \tau_0 I, \tag{5.8}$$

where the factor of proportionality, τ_0, is the minority carrier lifetime. The diffusion capacitance, C_d, can be written as

$$C_d \equiv \frac{dQ_d}{dV_j} = \frac{dQ_d}{dI} \cdot \frac{dI}{dV_j} = \tau_0 \frac{dI}{dV_j} \tag{5.9}$$

which by utilizing Equation (5.1) results in

$$C_d = \frac{\tau_0 I_0}{V_T} e^{V_j/V_T} \tag{5.10}$$

or in

$$C_d = \tau_0 \frac{I + I_0}{V_T}. \tag{5.11}$$

* The following discussion assumes that the two charges are each stored in a single storage element—the simplest possible assumption. For more detailed models see Linvill in references.

The charge Q_t and the capacitance C_t are functions of the effective width of the junction, which is a function of V_j. Neglecting stray capacitances, the second term of Equation (5.6) is the transition capacitance, C_t, that can be written as

$$C_t \equiv \frac{dQ_t}{dV_j} = \frac{C_0}{\left(1 - \frac{V_j}{\mathrm{V}_0}\right)^m}, \tag{5.12}$$

also, it can be shown that

$$Q_t = \frac{C_0 \mathrm{V}_0^m}{1 - m} [\mathrm{V}_0^{(1-m)} - (\mathrm{V}_0 - V_j)^{(1-m)}], \tag{5.13}$$

where m is between $\frac{1}{3}$ and $\frac{1}{2}$, and for silicon diodes V_0 is in the vicinity of 0.7 volts.*

BODY RESISTANCE AND CONDUCTIVITY MODULATION

The semiconductor materials in the diode are of finite size and result in a series ohmic body resistance which is approximately constant at low currents. At high currents, however, the stored charge may substantially increase the conductivity of the body material: this is the phenomenon of conductivity modulation. Although the exact calculations of diode conductance are complex, the conductance $1/r_s$ can be approximated by the sum of two terms. One of these is a constant g_0 which is characteristic of the structure, and the other one is proportional to the stored charge Q_d:

$$\frac{1}{r_s} = g_0 + \text{constant} \times Q_d. \tag{5.14}$$

Utilizing Equation (5.8), this can be written as

$$\frac{1}{r_s} = g_0 + \frac{I}{\mathrm{V}_s} \tag{5.15}$$

* Equations (5.12) and (5.13) are related by $C_t = dQ_t/dV_j$.

where the constant of proportionality, $1/V_s$, includes τ_0 of Equation (5.8). Substituting Equation (5.1) for I, Equation (5.15) can be also written as

$$\frac{1}{r_s} = g_0 + \frac{I_0}{V_s}(e^{V_j/V_T} - 1). \tag{5.16}$$

THE DIODE MODEL

The diode model (or equivalent circuit) incorporating the diode characteristics discussed above is shown in Fig. 5.3. For convenience, the expressions for I, r_i, C_d, C_t, and r_s are collected here:

$$I = I_0(e^{V_j/V_T} - 1) \tag{5.17a}$$

$$V_j = V_T \ln\left(1 + \frac{I}{I_0}\right), \tag{5.17b}$$

$$r_i = \frac{V_T}{I + I_0}, \tag{5.18}$$

$$C_d = \frac{\tau_0 I_0}{V_T} e^{V_j/V_T} = \tau_0 \frac{I + I_0}{V_T}, \tag{5.19}$$

$$C_t = \frac{C_0}{\left(1 - \frac{V_j}{V_0}\right)^m}, \tag{5.20}$$

$$\frac{1}{r_s} = g_0 + \frac{I}{V_s} = g_0 + \frac{I_0}{V_s}(e^{V_j/V_T} - 1). \tag{5.21}$$

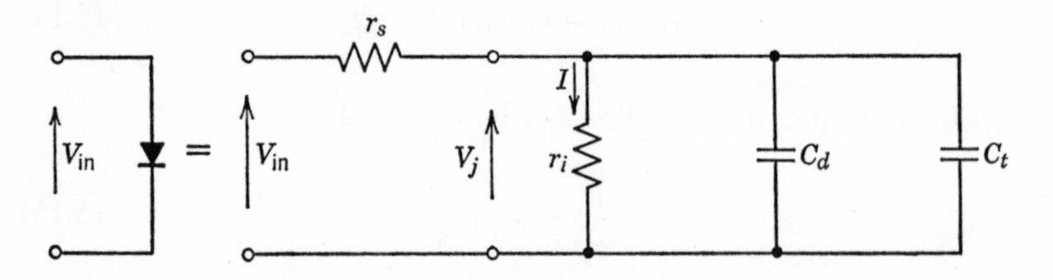

Figure 5.3 Junction diode model.

TRANSIENTS IN JUNCTION DIODE CIRCUITS

Transients of junction diodes in the circuit of Fig. 5.4(a) will be analyzed; the circuit with the diode model of Fig. 5.3 substituted is shown in Fig. 5.4(b) and the input generator current waveform $I_g(t)$ is shown in Fig. 5.4(c). Time t_{off} is chosen large enough ($6\,\tau_0$ in the examples) to attain equilibrium conditions; thus the turn-on transient at $t = 0$ and the turn-off transient at $t = t_{\text{off}}$ are separated.

Several special cases will be considered. First, the simplest case when C_t and r_s are zero and $R_g \to \infty$ will be analyzed. The analysis

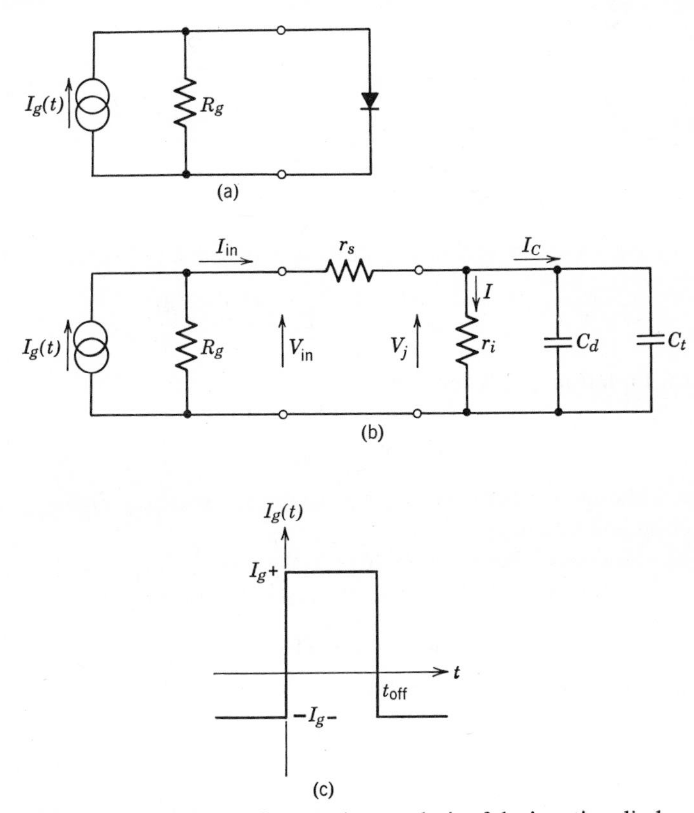

Figure 5.4 (a) Circuit for transient analysis of the junction diode. (b) The circuit with the model of Fig. 5.3 incorporated. (c) Generator current waveform ($t_{\text{off}} \gg \tau_0$).

will then be extended to the case of $r_s \neq 0$. Finally, results for finite R_g and for $C_t \neq 0$ will be given.

Transients with $C_t = 0$, $r_s = 0$, and $R_g \to \infty$

Here $V_{\text{in}} = V_j$, and resistor R_g is chosen to be arbitrarily large, hence, its current can be neglected.* Thus, the following equations can be written:

$$I_g = I + I_C, \tag{5.22}$$

$$I_C = C_d \frac{dV_j}{dt}, \tag{5.23}$$

and

$$I = I_0(e^{V_j/V_T} - 1) \tag{5.24a}$$

that can also be expressed as

$$V_j = V_T \ln\left(1 + \frac{I}{I_0}\right). \tag{5.24b}$$

Expanding Equation (5.23) and utilizing Equations (5.4) and (5.11),

$$I_C = C_d \frac{dV_j}{dt} = C_d \frac{dV_j}{dI}\frac{dI}{dt} = C_d r_i \frac{dI}{dt} = \tau_0 \frac{dI}{dt}; \tag{5.25}$$

hence, Equation (5.22) becomes

$$I_g = I + \tau_0 \frac{dI}{dt}. \tag{5.26}$$

Thus, although neither r_i nor C_d is constant, the resulting differential equation is linear in I.

The solution of Equation (5.26) for the $I_g(t)$ of Fig. 5.4(c) with $-I_{g-} < -I_0$ results in the turn-on transient diode current:

$$I = -I_0 + (I_{g+} + I_0)(1 - e^{-t/\tau_0}). \tag{5.27}$$

Also, by utilizing Equation (5.24b),

$$V_j = V_T \ln\left[\left(\frac{I_{g+}}{I_0} + 1\right)(1 - e^{-t/\tau_0})\right]. \tag{5.28}$$

* Resistor R_g cannot, however, be entirely omitted, since then all of $I_g(t)$ would be forced into the junction, and Equation (5.1) would be violated when $I_g(t)$ is more negative than $-I_0$.

When $I_{g+} \gg I_0$, Equation (5.28) can be approximated as

$$V_j \approx \mathrm{V_T} \ln \frac{I_{g+}}{I_0} + \mathrm{V_T} \ln (1 - \mathrm{e}^{-t/\tau_0}). \qquad (5.29)$$

The turn-off transient, assuming that equilibrium conditions are established by the time t_{off}, is given by

$$I = I_{g+} - (I_{g+} + I_{g-})(1 - \mathrm{e}^{-(t-t_{\text{off}})/\tau_0}), \qquad (5.30)$$

valid as long as $I > -I_0$, i.e., until

$$t - t_{\text{off}} < \tau_0 \ln \frac{I_{g+} + I_{g-}}{I_{g-} - I_0} ;$$

after this time $I \simeq -I_0$.

The corresponding voltage, V_j, can be obtained by substituting Equation (5.30) into Equation (5.24b):

$$V_j = \mathrm{V_T} \ln \left\{1 + \frac{1}{I_0} [-I_{g-} + (I_{g+} + I_{g-})\mathrm{e}^{-(t-t_{\text{off}})/\tau_0}]\right\}. \qquad (5.31)$$

When $I \gg I_0$, Equation (5.31) can be approximated as

$$V_j \approx \mathrm{V_T} \ln \frac{I_{g+}}{I_0} + \mathrm{V_T} \ln \left[- \frac{I_{g-}}{I_{g+}} + \left(1 + \frac{I_{g-}}{I_{g+}}\right) \mathrm{e}^{-(t-t_{\text{off}})/\tau_0} \right]. \qquad (5.32)$$

Representative voltage transients plotted from Equations (5.28) and (5.31) are shown in Fig. 5.5.* It is seen that for forward currents of $I_{g+} \gg I_0$, a change in I_0 may shift the voltage V_j but it does not change the shape of the transient [see also Equations (5.29) and (5.32)].

The stored charge and the voltage V_j become zero at a time $t_{\text{off}} + t_s$, where t_s is called the storage time of the diode. From Equation (5.31), by setting $V_j = 0$, the storage time is

$$t_s = \tau_0 \ln \left(1 + \frac{I_{g+}}{I_{g-}}\right). \qquad (5.33)$$

* An $I_{g+}/\mathrm{e}^{20} I_0 = 1$ corresponds to $I_{g+} \approx 5 \times 10^8 I_0$, e.g., to an $I_{g+} \approx 5$ mA for an $I_0 = 10$ nA.

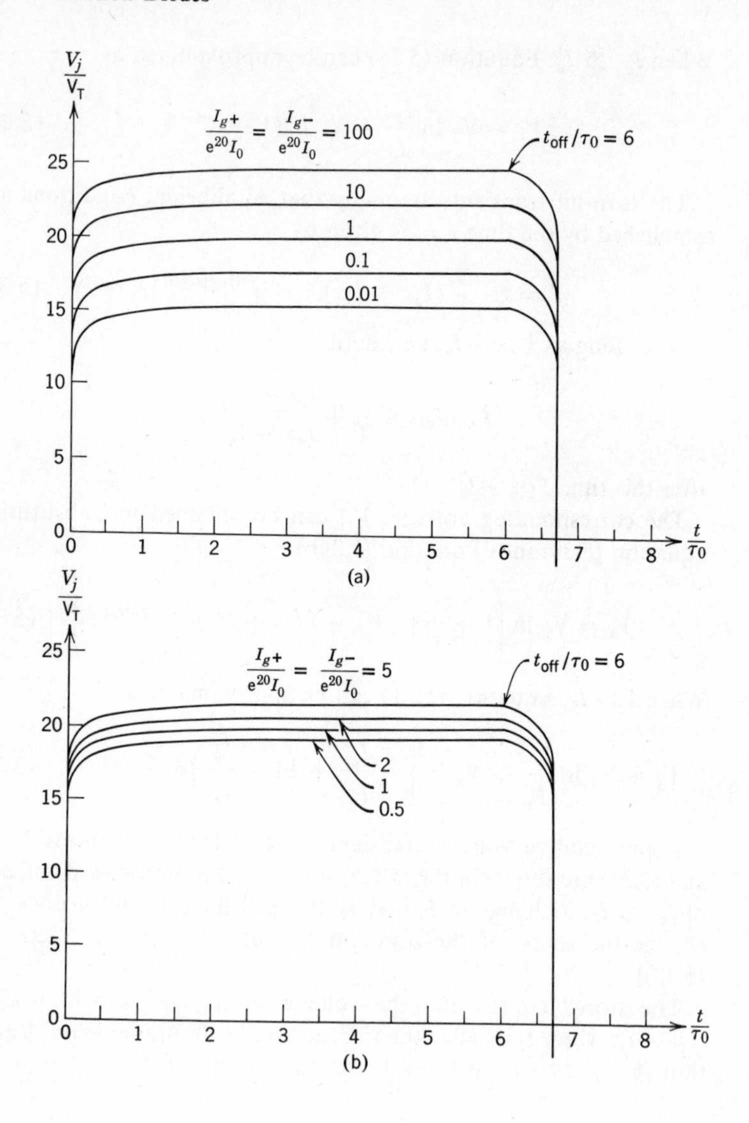
$\frac{V_j}{V_T}$
$\frac{I_g+}{e^{20}I_0} = \frac{I_g-}{e^{20}I_0} = 100$
$t_{off}/\tau_0 = 6$
10
1
0.1
0.01
25
20
15
10
5
0
0 1 2 3 4 5 6 7 8
$\frac{t}{\tau_0}$
(a)
$\frac{V_j}{V_T}$
$\frac{I_g+}{e^{20}I_0} = \frac{I_g-}{e^{20}I_0} = 5$
$t_{off}/\tau_0 = 6$
2
1
0.5
25
20
15
10
5
0
0 1 2 3 4 5 6 7 8
$\frac{t}{\tau_0}$
(b)

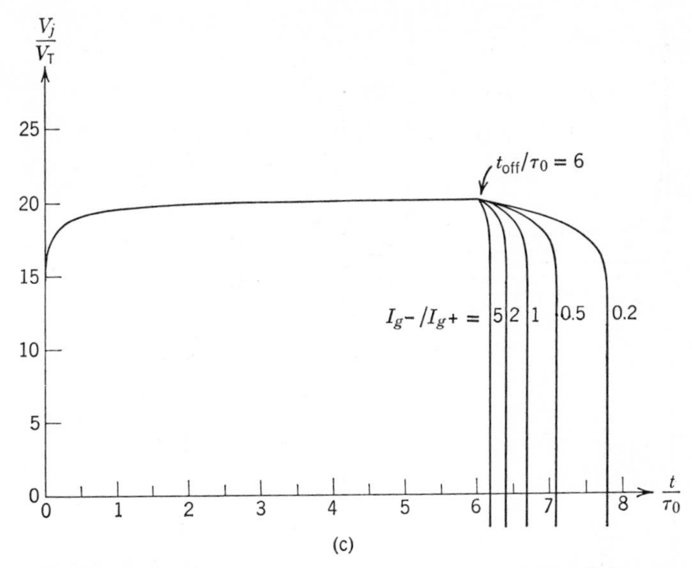

Figure 5.5 Transient response of a junction diode with $C_t = 0$, $r_s = 0$, and $R_g \to \infty$: (a) and (b) $I_{g+} = I_{g-}$ and with $\frac{I_{g+}}{e^{20}I_0} = \frac{I_{g-}}{e^{20}I_0}$ as parameter. (c) $\frac{I_{g+}}{e^{20}I_0} = 1$ and with $\frac{I_{g-}}{I_{g+}}$ as parameter.

Transients with $C_t = 0,\ r_s \neq 0,\ R_g \to \infty$

This case shows the effects of conductivity modulation. Here the diode input voltage can be written as

$$V_{\text{in}} = V_j + \frac{I_{\text{in}}}{g_0 + \dfrac{I}{\text{V}_s}}. \tag{5.34}$$

For $t < t_{\text{off}}$ and $I \gg I_0$, the turn-on transient can be obtained by utilizing Equations (5.27) and (5.29) for I and V_j, respectively. The input voltage thus resulting is

$$V_{\text{in}} = \text{V}_\text{T}\left[\ln\frac{I_{g+}}{I_0} + \ln\left(1 - e^{-t/\tau_0}\right) + \frac{\text{V}_\text{s}/\text{V}_\text{T}}{\dfrac{g_0\text{V}_\text{s}}{I_{g+}} + 1 - e^{-t/\tau_0}}\right]. \tag{5.35}$$

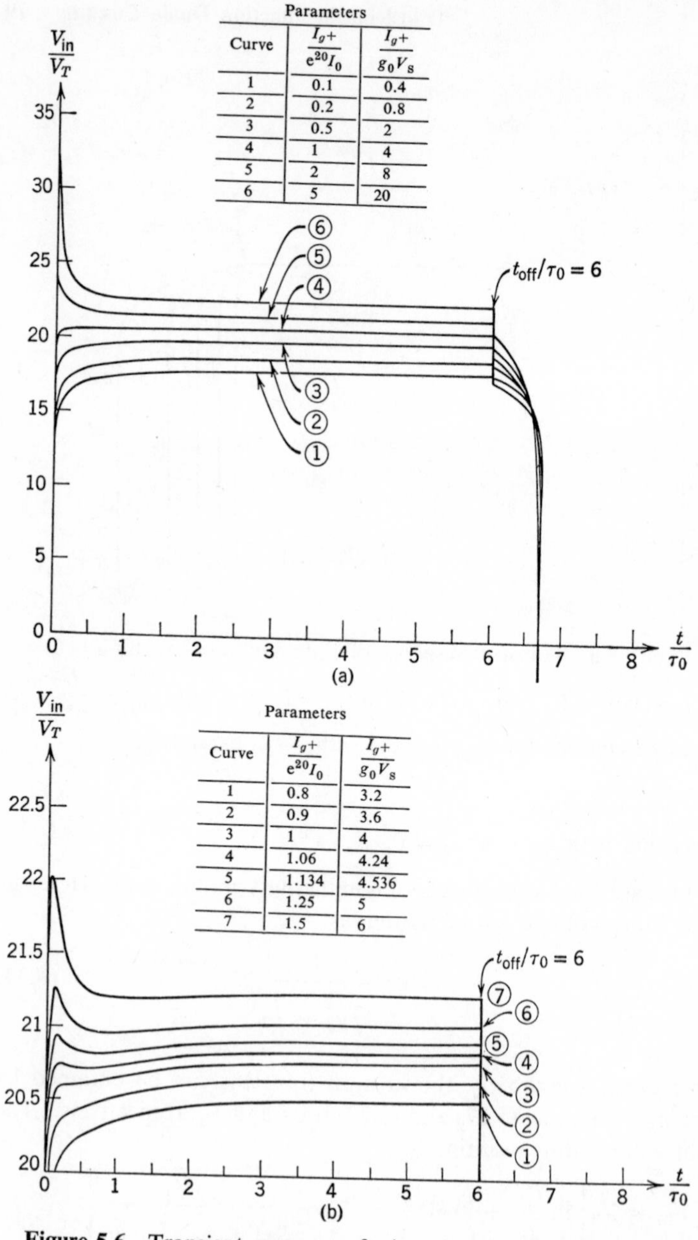

Parameters (a)

Curve	$\frac{I_{g+}}{e^{20}I_0}$	$\frac{I_{g+}}{g_0V_s}$
1	0.1	0.4
2	0.2	0.8
3	0.5	2
4	1	4
5	2	8
6	5	20

Parameters (b)

Curve	$\frac{I_{g+}}{e^{20}I_0}$	$\frac{I_{g+}}{g_0V_s}$
1	0.8	3.2
2	0.9	3.6
3	1	4
4	1.06	4.24
5	1.134	4.536
6	1.25	5
7	1.5	6

Figure 5.6 Transient response of a junction diode with $C_t = 0$, $r_s \neq 0$, $R_g \to \infty$, $g_0V_s = \frac{e^{20}I_0}{4}$, and $I_{g+} = I_{g-}$. Note different voltage scales and parameters in (a) and in (b).

It can be shown that the nature of the turn-on voltage transient is determined by the value of I_{g+}/g_0V_s. When $I_{g+}/g_0V_s < 4$, the turn-on transient is monotonic. When $4 < I_{g+}/g_0V_s < 4.536$, the turn-on transient has a local maximum followed by a local minimum, but the value of the maximum is less than the final steady state value of the voltage. For $I_{g+}/g_0V_s > 4.536$, the transient has an overshoot followed by an undershoot. As I_{g+}/g_0V_s is increased further above 4.536, the magnitude of the overshoot increases, while that of the undershoot decreases. These various possibilities are illustrated in Fig. 5.6.

Figure 5.6 also shows the turn-off transient obtained by substituting Equations (5.30) and (5.31) into Equation (5.34). As a result of nonzero r_s, there is an instantaneous voltage drop as the current reverses, followed by a storage time similar to that of the case of $r_s = 0$.

Transients with $R_g \neq \infty$, $C_t \neq 0$

If R_g is finite or C_t is not zero, the analysis of the transient response of the circuit of Fig. 5.4(b) becomes considerably more complicated, and the use of a digital computer becomes necessary. The following equations can be written:

$$C_d = \frac{\tau_0 I_0}{V_T} e^{V_j/V_T} \tag{5.36}$$

$$C_t = \frac{C_0}{\left(1 - \dfrac{V_j}{V_0}\right)^m} \tag{5.37}$$

$$I = I_0(e^{V_j/V_T} - 1) \tag{5.38}$$

$$r_s = \frac{1}{g_0 + I/V_s} \tag{5.39}$$

$$I_{in} = \frac{I_g R_g - V_j}{R_g + r_s} \tag{5.40}$$

$$V_{in} = (I_g - I_{in})R_g \tag{5.41}$$

$$I_C = I_{in} - I. \tag{5.42}$$

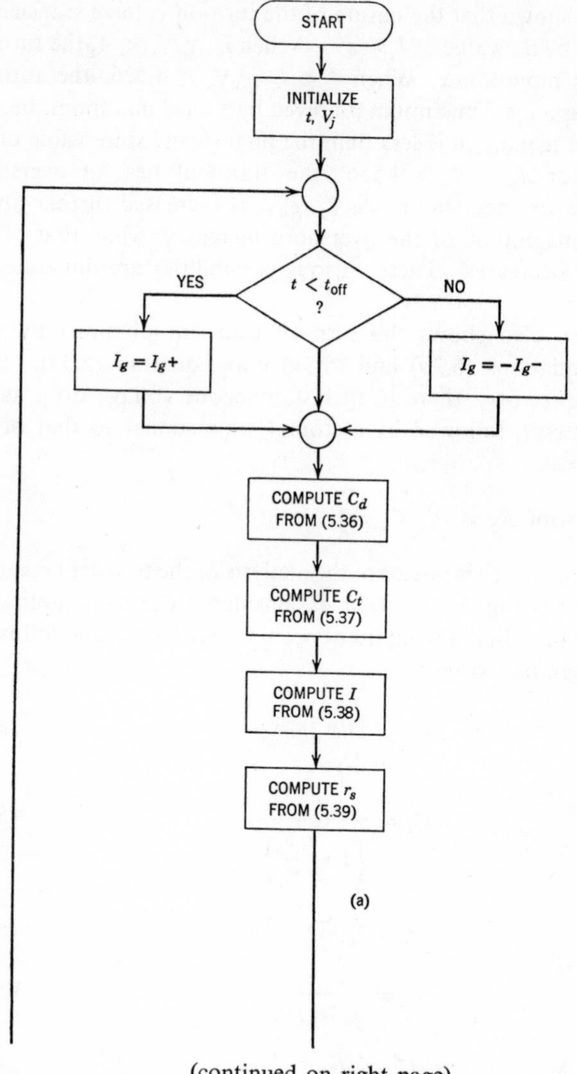

(continued on right page)

Figure 5.7 (a) and (b) Flowchart of the computer program.

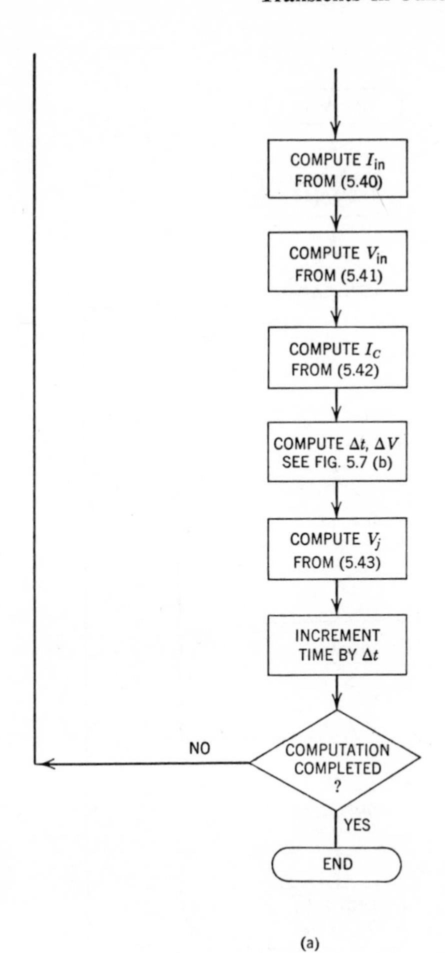

(continued from left page)

Figure 5.7(a) (Continued).

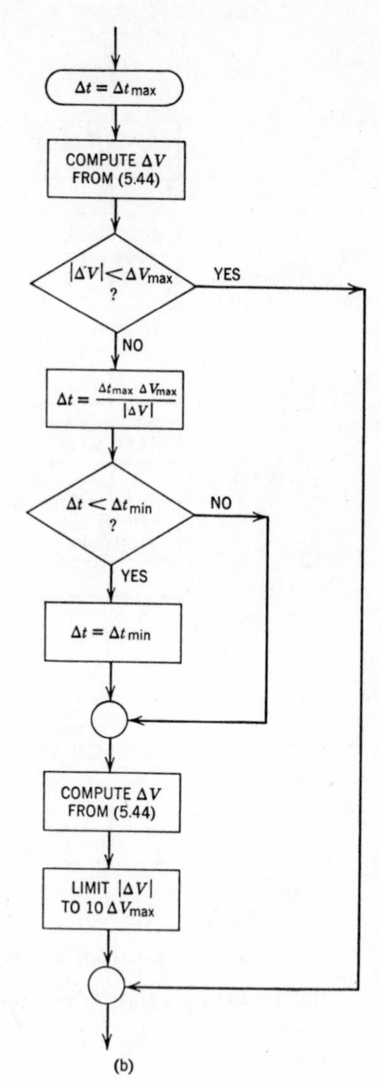

$\Delta t = \Delta t_{max}$
COMPUTE ΔV FROM (5.44)
$|\Delta V| < \Delta V_{max}$?
YES
NO
$\Delta t = \frac{\Delta t_{max}\ \Delta V_{max}}{|\Delta V|}$
$\Delta t < \Delta t_{min}$?
NO
YES
$\Delta t = \Delta t_{min}$
COMPUTE ΔV FROM (5.44)
LIMIT $|\Delta V|$ TO $10\,\Delta V_{max}$
(b)

Figure 5.7(b)

In addition,

$$V_j = \int \frac{I_C}{C_d + C_t}\, dt,$$

which integral can be approximated by a finite sum written as

$$V_j(t + \Delta t) = V_j(t) + \Delta V \tag{5.43}$$

where

$$\Delta V \equiv \frac{I_C\, \Delta t}{C_d + C_t}\,. \tag{5.44}$$

Equations (5.36) through (5.44) have been solved by using a digital computer. The flowchart of the program is shown in Fig. 5.7; $\Delta t_{\max} = 0.1\ \tau_0$, $\Delta t_{\min} = 10^{-6}\ \tau_0$, and $\Delta V_{\max} = 0.1\ \mathrm{V_T}$ have been used. The Fortran-IV computer program is shown in Fig. 5.8.

Representative voltage transients for $R_g \neq \infty$ are shown in Fig. 5.9 and Fig. 5.10; storage times are summarized in Fig. 5.11. Voltage transients for $C_t \neq 0$ with $\mathrm{V_0} = 25\ \mathrm{V_T}$ and $m = \frac{1}{2}$ are shown in Fig. 5.12 and Fig. 5.13.

Representative voltage transients for $R_g \neq \infty$ with zero and with finite body resistances are shown in Fig. 5.9 and Fig. 5.10, respectively. Storage times are summarized in Fig. 5.11. In the case when source resistance R_g is infinite, the storage time is given by Equation (5.33) as

$$t_s = \tau_0 \ln\left(1 + \frac{I_{g+}}{I_{g-}}\right). \tag{5.45}$$

Furthermore, when $I_{g-} \gg I_{g+}$, Equation (5.45) can be approximated, by utilizing $\ln(1 + x) \approx x$ for $|x| \ll 1$, as $t_s \approx \tau_0 I_{g+}/I_{g-}$, as shown by the broken line in Fig. 5.11. It can be also seen that there is some improvement in the storage time for finite source resistance R_g.

Transients for finite transition capacitances are illustrated in Fig. 5.12 and Fig. 5.13 with zero and finite body resistances, respectively. As expected, these show deteriorations in the risetime and in the fall time resulting from the finite transition capacitance.

```
C      *** DIODE TRANSIENT***
C
       FUNCTION CQ(V)
       IF(V.LT.-100.0)V=-100.0
       IF(V.GT.+140.0)V=+140.0
       CQ=EXP(V-20.0)
       RETURN
       END
C
       FUNCTION DIODE(V)
       IF(V.LT.-100.0)V=-100.0
       IF(V.GT.+120.0)V=+120.0
       DIODE=EXP(V)-1.0
       RETURN
       END
C
       REAL ID,IC,IG,L
       DOUBLE PRECISION V
       CALL STRTP1(29)
       CALL PLOT1(0.0,-30.0,23)
       CALL PLOT1(0.0,0.5,23)
     1 FORMAT(6F10.4,I1)
     2 FORMAT(' ','RISETIME=',1PE10.3,'  TURNOFFTIME=',1PE10.3)
     3 FORMAT(' ',7F10.4,I1)
     4 FORMAT('1')
     5 FORMAT(' ')
     6 FORMAT(F10.4)
       ARG1=-20.0
       EXP1=EXP(ARG1)
     7 CONTINUE
       READ(5,6)CTR
       IF(CTR.LE.0.0)GO TO 100
       IF(CTR.LT.0.001)CTR=0.0
       WRITE(6,4)
    11 CONTINUE
       READ(5,1)G,GN,R1,DELT,L,C1,NEWPLT
       IF(G.LE.0.0)GO TO 7
       WRITE(6,5)
       WRITE(6,3)G,GN,R1,DELT,L,C1,CTR,NEWPLT
       IF(C1.GT.100.0)C1=1E20
       DELT2=1E-6
       CALL PLOT1(0.0,0.0,+3)
       IF(NEWPLT.EQ.0)GO TO 12
       CALL PLOT1(15.0,0.0,-3)
       CALL AXIS1(0.0,2.0,'T',-1,10.0,0.0,0.0,1.0,20.0)
       CALL AXIS1(0.0,0.0,'V',+1,10.0,90.0,-10.0,5.0,10.0)
    12 CONTINUE
       TRISE=0.0
       TTOFF=0.0
       XPLOTM=-0.02
       V=-GN
       VL=-GN
       IC=0.0
```

Figure 5.8 Fortran-IV program.

```
      T=0.0
      XPLOT=0.0
      YPLOT=2.0+0.2*VL
      IF(YPLOT.LE.0.0)YPLOT=0.0
      CALL PLOT1(XPLOT,YPLOT,+3)
   13 CONTINUE
      V1=V
      IF(CTR.LE.0.0)C=CQ(V1)
      IF(CTR.GT.0.0)C=CQ(V1)+CTR/SQRT(1.0-V1*0.04)
      IF(T.LT.6.0)VG=G
      IF(T.GE.6.0)VG=-GN
      ID=DIODE(V)*EXP1
      GS=C1+L*ID
      GS1=1.0/GS
      IG=(VG-V)/(GS1+R1)
      VL=VG-IG*R1
      IC=IG-ID
      DELT1=DELT
      DV=IC*DELT1/C
      ABSDV=ABS(DV)
      IF(ABSDV.LT.0.1)GO TO 16
      DELT1=0.1*DELT1/ABSDV
      IF(DELT1.LT.DELT2)DELT1=DELT2
      DV=IC*DELT1/C
      IF(DV.GT.0.1)DV=0.1
      IF(DV.LT.-0.1)DV=-0.1
   16 CONTINUE
      V=V+DV
      T=T+DELT1
      IF(VL.GT.0.0.AND.TRISE.EQ.0.0)TRISE=T
      IF(VL.LT.0.0.AND.TRISE.NE.0.0.AND.TTOFF.EQ.0.0)TTOFF=T-6.0
      XPLOT=T
      IF(XPLOT-XPLOTM.LT.0.01)GO TO 15
      XPLOTM=XPLOT
      YPLOT=2.0+0.2*VL
      IF(YPLOT.LT.0.0)YPLOT=0.0
      IF(YPLOT.GT.10.0)YPLOT=10.0
   14 CONTINUE
      CALL PLOT1(XPLOT,YPLOT,+2)
      IF(YPLOT.LE.0.0.AND.T.GT.6.0)GO TO 17
   15 CONTINUE
      IF(XPLOT.LE.10.)GO TO 13
   17 CONTINUE
      WRITE(6,2)TRISE,TTOFF
      WRITE(6,5)
      GO TO 11
  100 CONTINUE
      CALL PLOT1(15.0,0.0,-3)
      CALL ENDP1
      STOP
      END
```

Figure 5.8 (Continued).

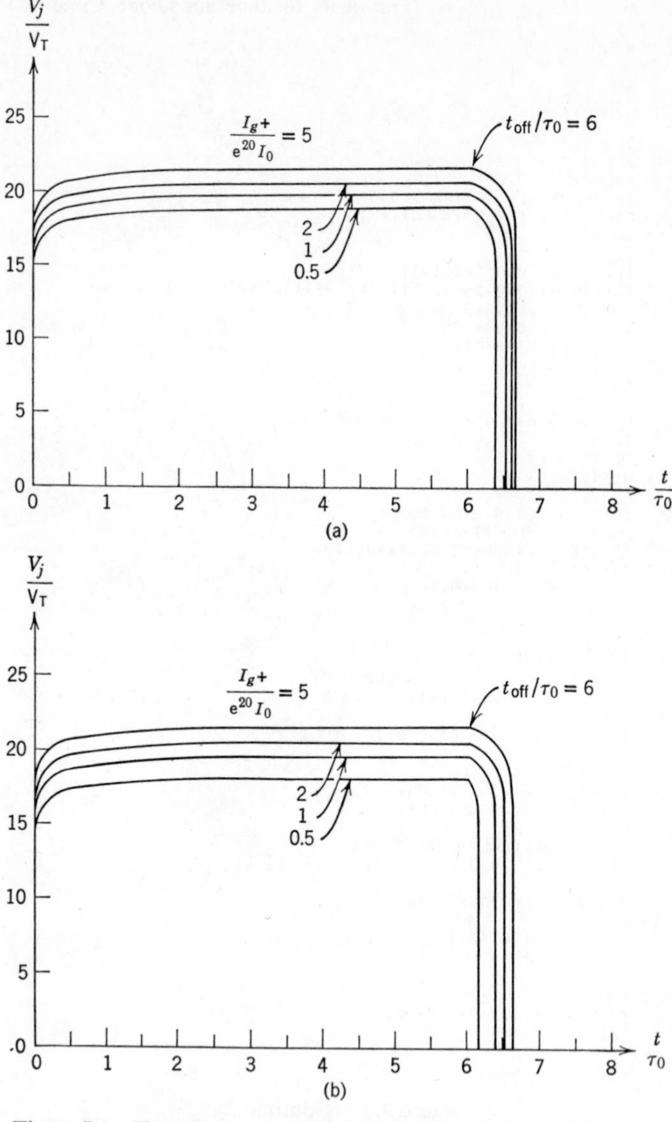

Figure 5.9 Transient response of a junction diode with $C_t = 0$, $R_g \neq \infty$, $r_s = 0$, $I_{g+} = I_{g-}$, and with $\frac{I_{g+}}{e^{20}I_0} = \frac{I_{g-}}{e^{20}I_0}$ as parameter: (a) $R_g\,e^{20}I_0/V_T = 100$. (b) $R_g\,e^{20}I_0/V_T = 50$.

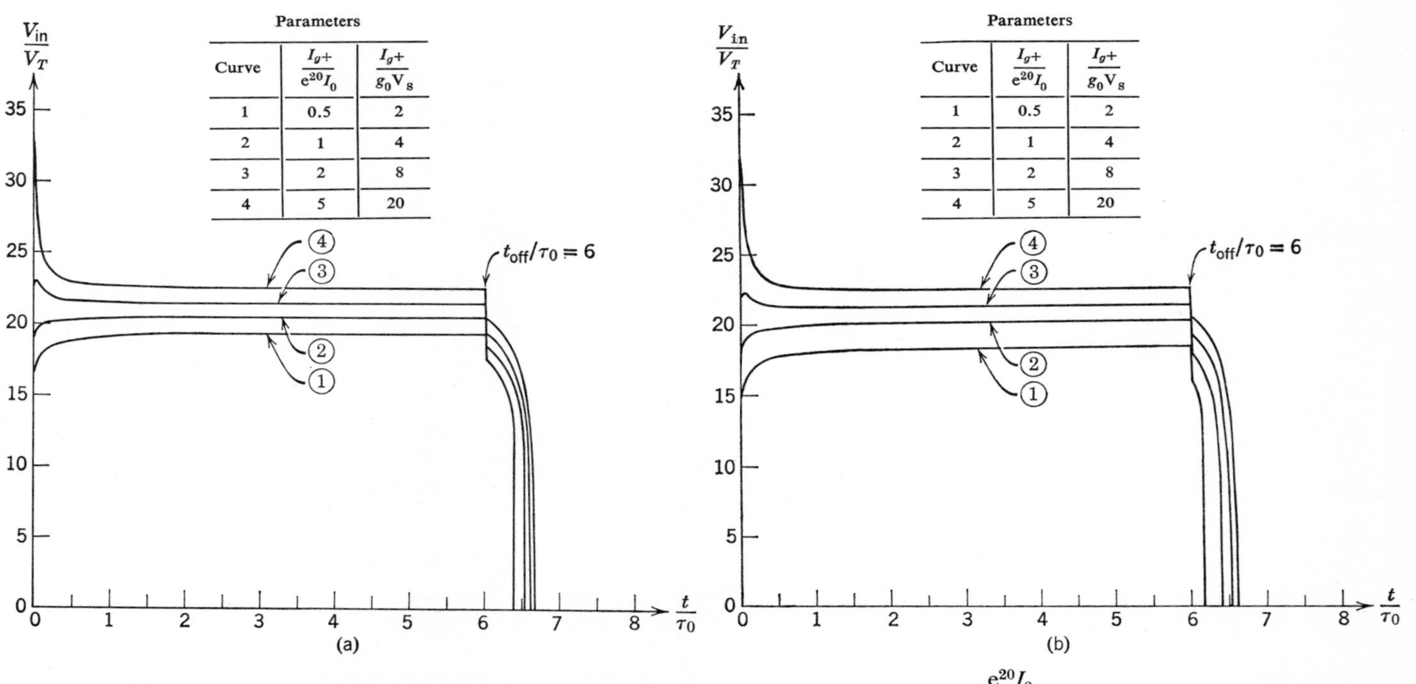

Figure 5.10 Transient response of a junction diode with $R_g \neq \infty$, $r_s \neq 0$, $g_0 V_s = \dfrac{e^{20} I_0}{4}$, and $I_{g+} = I_{g-}$: (a) $R_g\, e^{20} I_0 / V_T = 100$. (b) $R_g\, e^{20} I_0 / V_T = 50$.

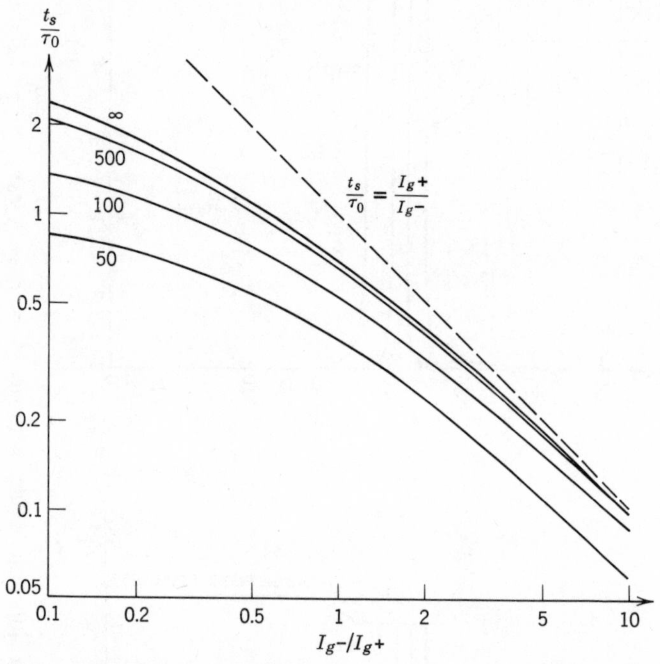

Figure 5.11 Storage times $t_s \equiv t(V = 0) - t_{\text{off}}$ as function of I_{g-}/I_{g+} for $I_{g+} = e^{20}I_0$ with normalized source resistance $R_g I_{g+}/V_T$ as parameter.

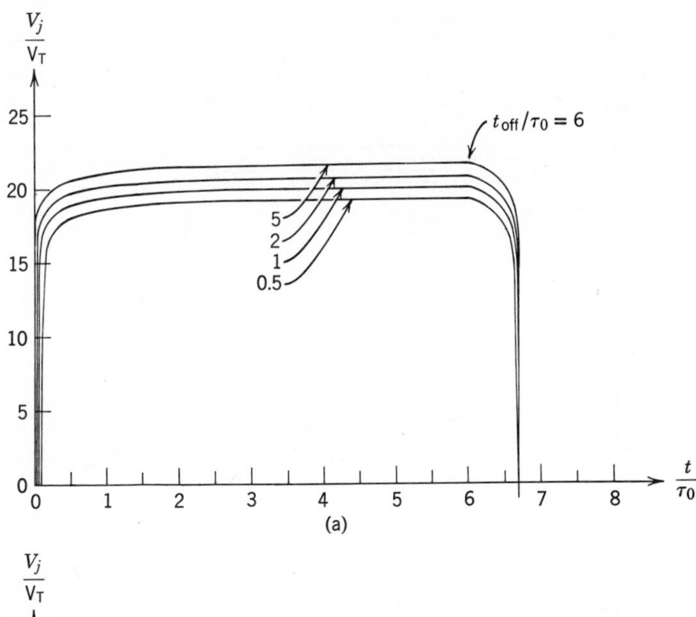

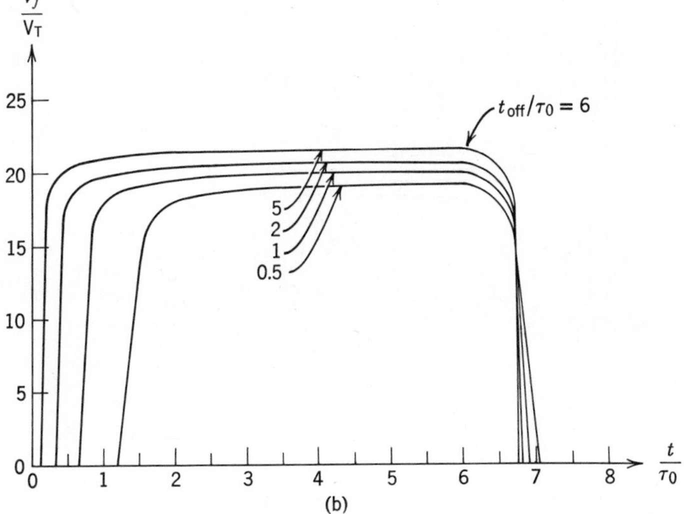

Figure 5.12 Transient response of a junction diode with $C_t \neq 0$, $R_g = \infty$, $r_s = 0$, $V_0 = 25\,V_T$, $m = 0.5$, $I_{g^+} = I_{g^-}$, and with $\dfrac{I_{g^+}}{e^{20} I_0}$ as parameter: (a) $\dfrac{C_0 V_T}{\tau_0 I_0 e^{20}} = 0.001$. (b) $\dfrac{C_0 V_T}{\tau_0 I_0 e^{20}} = 0.01$.

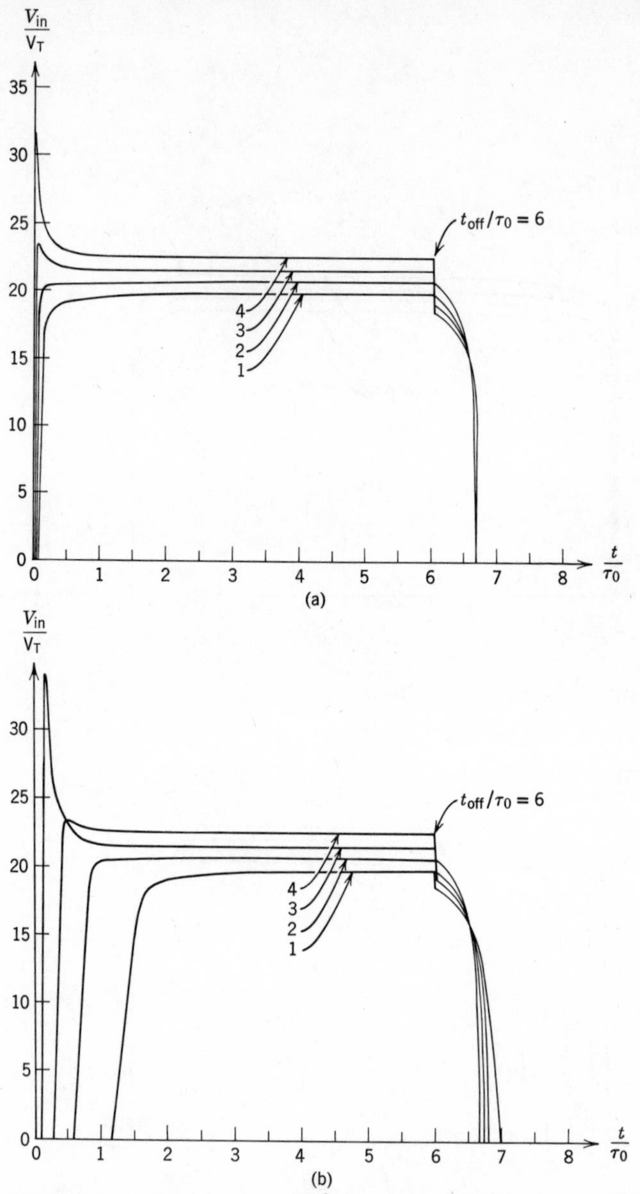

Figure 5.13 Transient response of a junction diode with $C_t \neq 0$, $R_g = \infty$, $r_s \neq 0$, $g_0 V_s = \dfrac{e^{20} I_0}{4}$, $V_0 = 25\ V_T$, $m = 0.5$, and $I_{g+} = I_{g-}$: (a) $\dfrac{C_0 V_T}{\tau_0 I_0 e^{20}} = 0.001$. (b) $\dfrac{C_0 V_T}{\tau_0 I_0 e^{20}} = 0.01$.

EXERCISES

1. Calculate the diode current I of Equation (5.1) for a diode with $I_0 = 10$ nA, $n = 1.5$, $kT/q = 25$ mV, and for $V_j = -10$ V, -1 V, 0 V, $+0.5$ V, and $+0.6$ V.

2. Utilize Equations (5.1) and (5.2) to show that the temperature coefficient of the junction voltage of a diode, operating at a constant current, at a $V_j = 0.6$ V, and at a temperature of 300°K is $dV_j/dT \approx -2$ mV/°C if

$$\frac{1}{I_0}\frac{dI_0}{dT} = 0.08/°\text{C}.$$

3. Determine V_1 in Fig. 5.14 if the diodes are identical, assuming $I_0 \ll 0.1\, I_1$ and $V_T = 30$ mV.

4. Calculate the incremental resistance r_i and the diffusion capacitance C_d for the diode of Exercise 1, given $\tau_0 = 100$ ps.

5. Show that $r_i C_d = \tau_0$.

6. Find the transition capacitance C_t for the diode of Exercise 1 with $C_0 = 1$ pF, $m = 0.5$, and $V_0 = 0.75$ volt.

7. Utilize Figure 5.11 to find the storage time of a diode with $I_0 = 10$ nA, $n = 1.5$, $kT/q = 25$ mV, $V_j = +0.6$ V, and $\tau_0 = 100$ ps, if $I_{g+} = I_{g-}$. Compute the storage time for $R_g = 20$ kΩ and for $R_g \to \infty$.

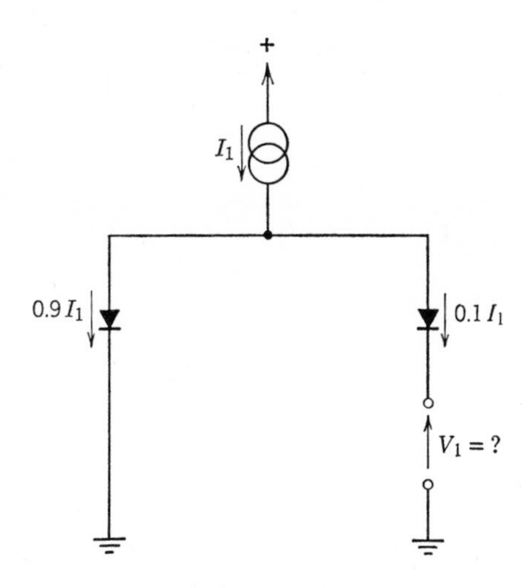

Figure 5.14

8. What is the maximum forward current that can be turned on from a current source into a diode with $C_t \approx 0$, $1/g_0 = 3\ \Omega$, and $V_s = V_T = 30$ mV, if a voltage transient with no overshoot is desired.

9. How large is the overshoot on the diode of Exercise 8 if the forward current is twice that computed there.

10. Sketch the approximate *current* waveforms of I_{in} in Fig. 5.4 for the transients of Fig. 5.9.

ooooo

6

ooooo

Tunnel-Diodes

oooooooooooooooooooooooooooooo

The tunnel-diode, which is widely used in high speed circuits, is a junction diode with relatively large impurity concentration near its junction resulting in high currents at near zero voltages. Consequently, the current versus voltage characteristics are considerably more involved than those of junction diodes. As a rough approximation (see Ferendeci and Ko in the references), the current I of the tunnel-diode can be approximated by a sum of a "junction-diode current" $I_0(e^{V/V_T} - 1) \approx I_0 e^{V/V_T}$ and a "tunnel-current" of the form $(V/V_p)e^{-V/V_p}$. One such approximation for germanium diodes is

$$I = I_p\left(\frac{V}{V_p}\, e \times e^{-V/V_p} + 5 \times 10^{-10} e^{3V/V_p}\right), \tag{6.1}$$

shown in Fig. 6.1. The peak voltage V_p is typically in the vicinity of 100 mV, peak current I_p can range from 50 μA to 200 mA.

An equivalent circuit (or model) neglecting series resistance is shown in Fig. 6.2. Voltage-variable resistor $r(V)$ represents the dc I-V characteristic of Fig. 6.1 and Equation (6.1), capacitance C_2 is the junction capacitance, L is the lead inductance, and C_1 is the sum of

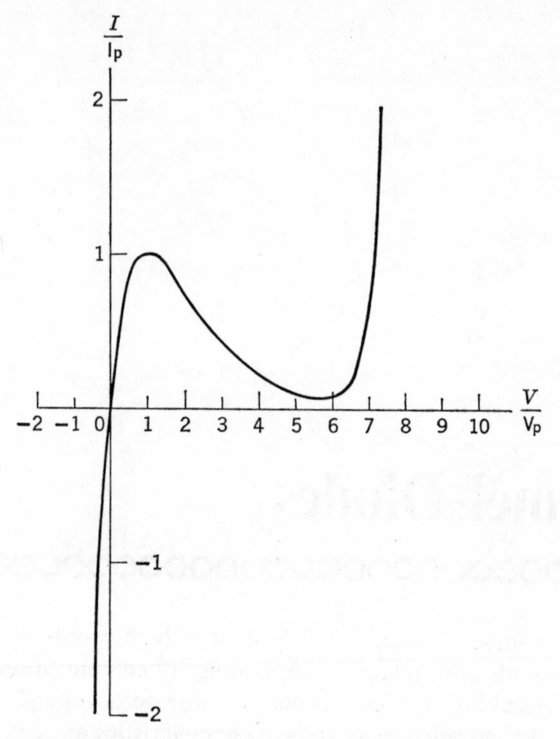

Figure 6.1 Tunnel-diode dc current versus voltage characteristic.

lead capacitances plus external capacitances. At low voltages, to a very good approximation, C_1, C_2, and L can be assumed to be independent of voltage and current; at voltages above $\approx 6\ V_p$ this assumption becomes inaccurate as a result of increasing diffusion capacitance [see Equations (5.10) and (5.11)].

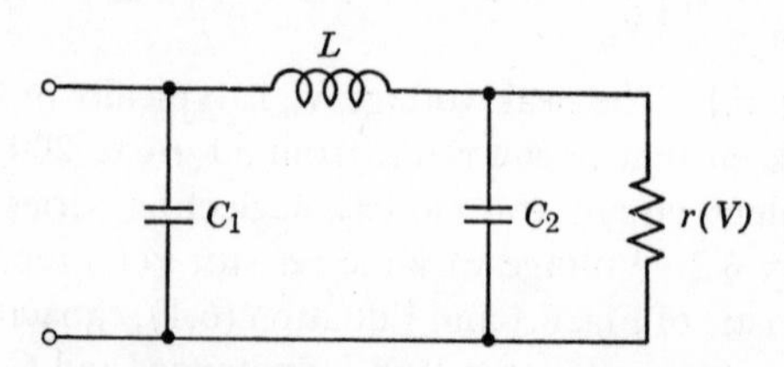

Figure 6.2 Tunnel-diode equivalent circuit.

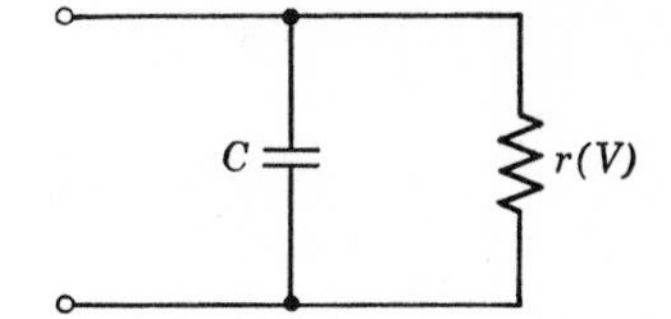

Figure 6.3 Simplified tunnel-diode equivalent circuit.

ANALYSIS OF TUNNEL-DIODE TRANSIENT USING A SIMPLIFIED EQUIVALENT CIRCUIT

In many cases, the equivalent circuit of Fig. 6.2 can be simplified to that of Fig. 6.3. This simplified equivalent circuit is utilized in the circuit of Fig. 6.4 to obtain some basic features of the tunnel-diode transient for a ramp-current input signal. Assuming a constant capacitance C and the dc current versus voltage characteristic of Equation (6.1), the following equations can be written:

$$V(t) = \int \frac{1}{C} I_C(t)\, dt \tag{6.2}$$

$$I_g(t) = kt \tag{6.3}$$

$$I_D(t) = \mathrm{I_p}\left[\frac{V(t)}{\mathrm{V_p}} \times \mathrm{e} \times \mathrm{e}^{-V(t)/\mathrm{V_p}} + 5 \times 10^{-10}\mathrm{e}^{3V(t)/\mathrm{V_p}}\right] \tag{6.4}$$

$$I_C(t) = I_g(t) - I_D(t) \tag{6.5}$$

and

$$V(t \leq 0) = 0. \tag{6.6}$$

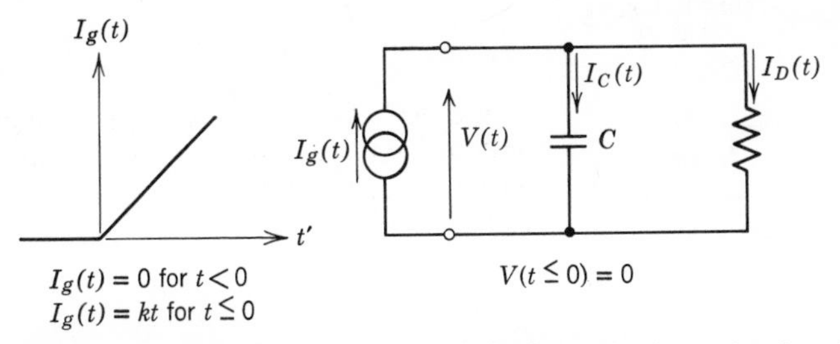

Figure 6.4 The simplified tunnel-diode equivalent circuit of Fig. 6.3 driven by a ramp-current input signal.

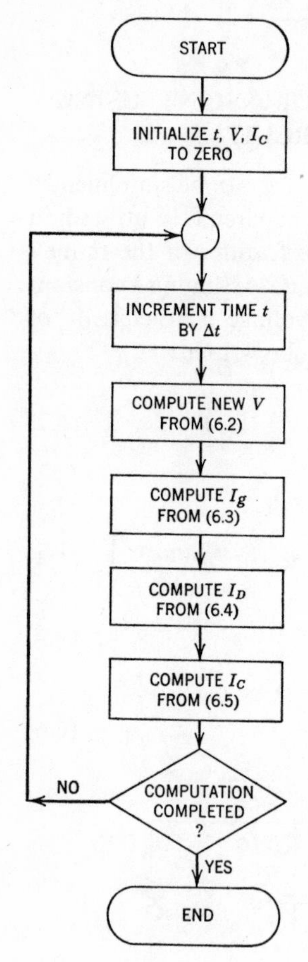

Figure 6.5 Flow-chart of the computer program for the transient computation of the circuit of Fig. 6.4.

```
      FUNCTION F1(Y,B1,B2)
      B2Y=B2*Y
      IF(B2Y.GT.150.0)B2Y=150.0
      IF(Y.LT.-100.0)Y=-100.0
      F1=Y*EXP(1-Y)  =-10*B1*EXP(B2Y)
      RETURN
      END
      CALL STRTP1(10)
      CALL PLOT1(20.0,0.0,-3)
      CALL AXIS1(0.0,1.0,'LOG T',-5,12.0,0.0,-1.0,0.25,2.5)
      CALL PLOT1(0.0,1.0,+3)
      CALL AXIS1(0.0,1.0,'LOG V',+5,8.0,90.0,-1.0,0.25,2.5)
    1 FORMAT(4F10.5)
      REAL I
   11 CONTINUE
      CALL PLOT1(0.0,1.0,+3)
      READ(5,1)A,DT,B1,B2
      IF(A.EQ.0)GO TO 100
      V=0
      I=0
      T=0
      J=0
      XPLOTM=-0.02
    2 CONTINUE
      J=J+1
      T=J*DT
      V=V+I*DT
      X=A*T
      I=X  -F1(V,B1,B2)
      IF(X.GE.100.0)GO TO 11
      IF(V.GE.10.0)GO TO 11
      IF(X.LE.0.1)GO TO 2
      IF(V.LE.0.1)GO TO 2
      XPLOT=4.0*ALOG10(10*X)
      IF(ABS(XPLOT-XPLOTM).LT.0.01)GO TO 3
      XPLOTM=XPLOT
      YPLOT=4.0*ALOG10(10*V)+1.0
      CALL PLOT1(XPLOT,YPLOT,2)
    3 CONTINUE
      GO TO 2
  100 CONTINUE
      CALL PLOT1(20.0,0.0,-3)
      CALL ENDP1
      STOP
      END
```

Figure 6.6 Fortran-IV program for the transient computation of the circuit of Fig. 6.4.

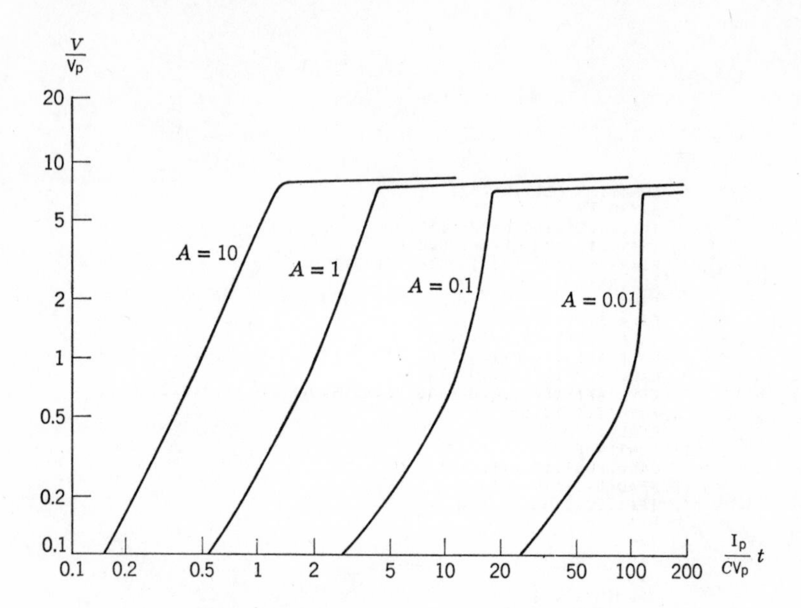

Figure 6.7 Transient response of the circuit of Fig. 6.4. Normalized voltage V/V_p is shown as a function of normalized time, $(\text{I}_\text{p}/C\text{V}_\text{p})t$, with normalized ramp-current slope $A \equiv kC\text{V}_\text{p}/\text{I}_\text{p}^2$ as parameter. *Note:* Both scales are logarithmic.

These equations are solved by converting Equation (6.2) into a finite sum and by using a digital computer. The flow-chart of the program is shown in Fig. 6.5 and the corresponding Fortran-IV program in Fig. 6.6.

Results for normalized ramp-current input slopes of $A \equiv kC\text{V}_\text{p}/\text{I}_\text{p}^2 = 0.01, 0.1, 1$, and 10 are shown in Fig. 6.7 using logarithmic time and voltage scales.* For fast-rising input currents (large A), the transient is dominated by capacitance C and the voltage is a quadratic function of time. For slow-rising input currents (small A), the voltage closely follows the dc characteristic for $V \leq \text{V}_\text{p}$, with a comparatively fast transition to $V > 6\ \text{V}_\text{p}$. This transition is shown in detail for $10^{-5} \leq A \leq 10$ in Fig. 6.8, using linear voltage scales

* In the computation, an incremental time of $(\text{I}_\text{p}/C\text{V}_\text{p})\Delta t = 0.0005$ was used for $A = 10$, 0.005 with other values of A.

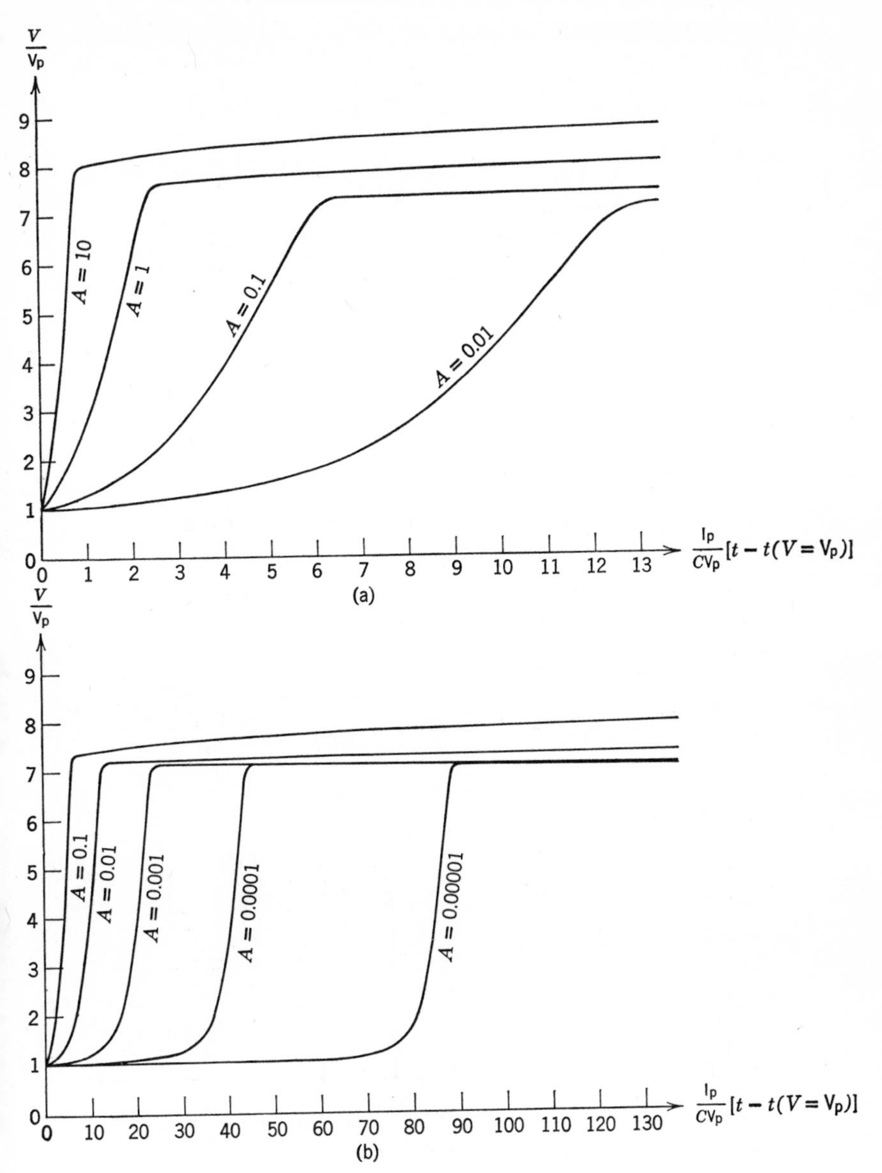

Figure 6.8 Normalized voltage V/V_p in the circuit of Fig. 6.4 as function of normalized time with the time scale offset by $t(V = V_p)$, and with the normalized ramp-current slope $A \equiv kCV_p/I_p^2$ as parameter.

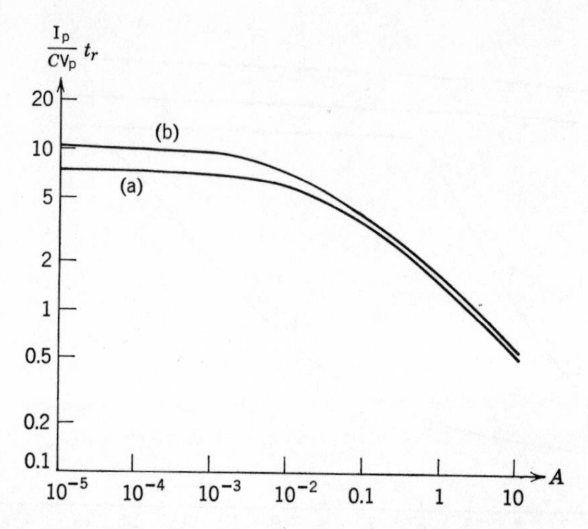

Figure 6.9 Summary of risetimes t_r from Fig. 6.8. (a) Risetime between 2 V_p and 7 V_p. (b) Risetime between 1.5 V_p and 6.5 V_p.

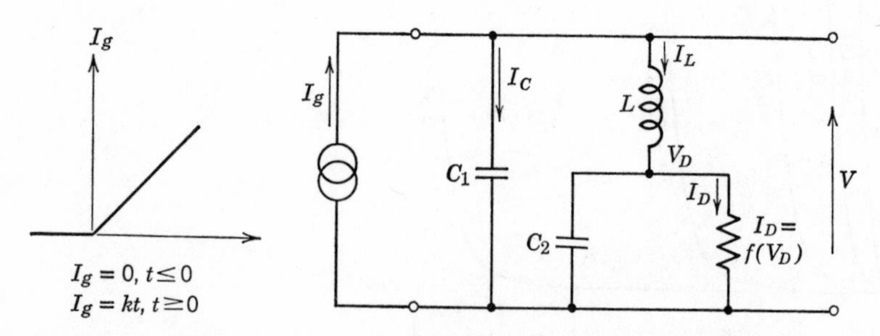

Figure 6.10 The complete tunnel-diode equivalent circuit of Fig. 6.2 driven by a ramp-current input.

and offset linear time scales. As the rate of rise of the input current, A, becomes smaller, there is more delay, but the shape of the transition becomes independent of A. This result can be also seen in Fig. 6.9, where risetimes are summarized for various values of A.

ANALYSIS OF TUNNEL-DIODE TRANSIENT USING THE COMPLETE EQUIVALENT CIRCUIT

The complete tunnel-diode equivalent circuit of Fig. 6.2 is utilized in the circuit of Fig. 6.10 to obtain a more exact solution of the tunnel-diode transient for a ramp-current input signal. Using Equation (6.1) for the dc characteristics, the following equations can be written:

$$I_C = kt - I_L \tag{6.7}$$

$$I_L = \int \frac{1}{L}(V - V_D)\,dt \tag{6.8}$$

$$V = \int \frac{1}{C_1} I_C\,dt \tag{6.9}$$

$$V_D = \int \frac{1}{C_2}(I_L - I_D)\,dt \tag{6.10}$$

$$I_D = \mathrm{I_p}\left[\frac{V_D}{\mathrm{V_p}}\,\mathrm{e}^{1-V_D/\mathrm{V_p}} + 5 \times 10^{-10}\mathrm{e}^{3V_D/\mathrm{V_p}}\right]. \tag{6.11}$$

These equations are solved by converting Equations (6.8), (6.9), and (6.10) into finite sums and by using a digital computer with the program flow-chart of Fig. 6.11 and the Fortran-IV program of Fig. 6.12. Representative results are shown in Fig. 6.13 for $C_2/C_1 = 0.05$, 0.1, and 0.2. Note that in some cases there is ringing on the top of the voltage waveform and/or on the rising edge. Since in many applications the tunnel-diode is utilized as a threshold element to provide a sharp transition when a selected input current is reached, it is advantageous to have a rising edge that is free of ringing. Examples of this are Fig. 6.13(c) and (d) with $A \geq 0.02$, and Fig. 6.13(e) and (f).

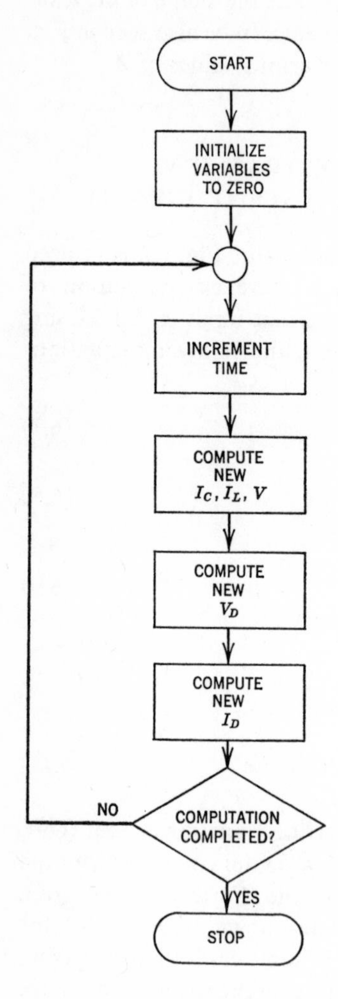

Figure 6.11 Flow-chart of the computer program for the transient computation of the circuit of Fig. 6.10.

```
      FUNCTION F1(Y,B1,B2)
      B2Y=B2*Y
      IF(B2Y.GT.150.0)B2Y=150.0
      IF(Y.LT.-100.0)Y=-100.0
      F1=Y*EXP(1-Y)+1E-10*B1*EXP(B2Y)
      RETURN
      END
 1    FORMAT(6F10.4,I5,F10.4,I5)
 20FORMAT('1','A=',F10.4,5X,'L=',F10.4,5X,'DX=',F10.4,5X,'B1=',F10.4,
  15X,'B2=',F10.4,5X,'C2=',F10.4)
 3 FORMAT('0',' ')
 40FORMAT(' ','X=',1PE9.2,' V2J=',1PE9.2,' V2JM1=',1PE9.2,' I2J=',
  11PE9.2,' I2JM1=',1PE9.2,' RATE=',1PE9.2,' V1J=',1PE9.2,' V1JM1=',
  21PE9.2,' ID=',1PE9.2)
 50FORMAT('0','XMAX=',1PE10.3,' FMAX=',1PE10.3,' XMIN=',1PE10.3,
  1' FMIN=',1PE10.3)
      REAL A,L,DX,B1,B2,X,V1J,I2JM1,I2J,V2JM1,V2J,V1JM1,I2JM2,ID
      CALL STRTP1(10)
 10   READ(5,1)A,L,DX,B1,B2,C2,M,V2JST,IRET
 11   IF(A.LE.0)GO TO 100
      CALL PLOT1(0.0,0.0,3)
      IF(IRET.EQ.0)GO TO 12
      CALL PLOT1(25.0,0.0,-3)
      CALL AXIS1(0.0,0.0,'T',-1,15.0,0.0,0.0,10.,10.0)
      CALL PLOT1(0.0,0.0,3)
      CALL AXIS1(0.0,0.0,'V',+1,10.,90.0,0.0,1.0,10.0)
      CALL PLOT1(0.0,0.0,3)
 12   WRITE(6,2)A,L,DX,B1,B2,C2
      WRITE(6,3)
      ID=0
      RATE=0
 13   X=0
 14   V1J=0
 15   I2JM1=0
 16   I2J=0
 17   V2JM1=0
 18   V2J=0
 19   X=X+DX
 20   V1JM1=V1J
 21   V1J=V1JM1+(A*X-I2JM1)*DX
 22   IF(V2J.GT.10*DX)RATE=(V2J-V2JM1)/(V2J*DX)
 23   I2JM1=I2J
 24   I2J=I2JM1+(V1JM1-V2JM1)*DX/L
 25   V2JM1=V2J
 26   V2J=V2JM1+(I2J-ID)*DX/C2
 27   ID=F1(V2J,B1,B2)
 28   Z=FLOAT(M)
 29   IF(ABS(INT(X/DX+DX)/Z-INT((X/DX+DX)/Z)).GE.DX)GO TO 31
      XPLOT=0.1*X
      YPLOT=V1J
      IF(XPLOT.LE.0.0)XPLOT=0.0
      IF(XPLOT.GE.15.)GO TO 10
      IF(YPLOT.LE.0.0)YPLOT=0.0
      IF(YPLOT.GE.10.)GO TO 10
      CALL PLOT1(XPLOT,YPLOT,2)
 31   CONTINUE
      GO TO 19
 100  CONTINUE
      CALL PLOT1(15.0,0.0,-23)
      CALL ENDP1
      STOP
      END
```

Figure 6.12 Fortran-IV program for the transient computation of the circuit of Fig. 6.10.

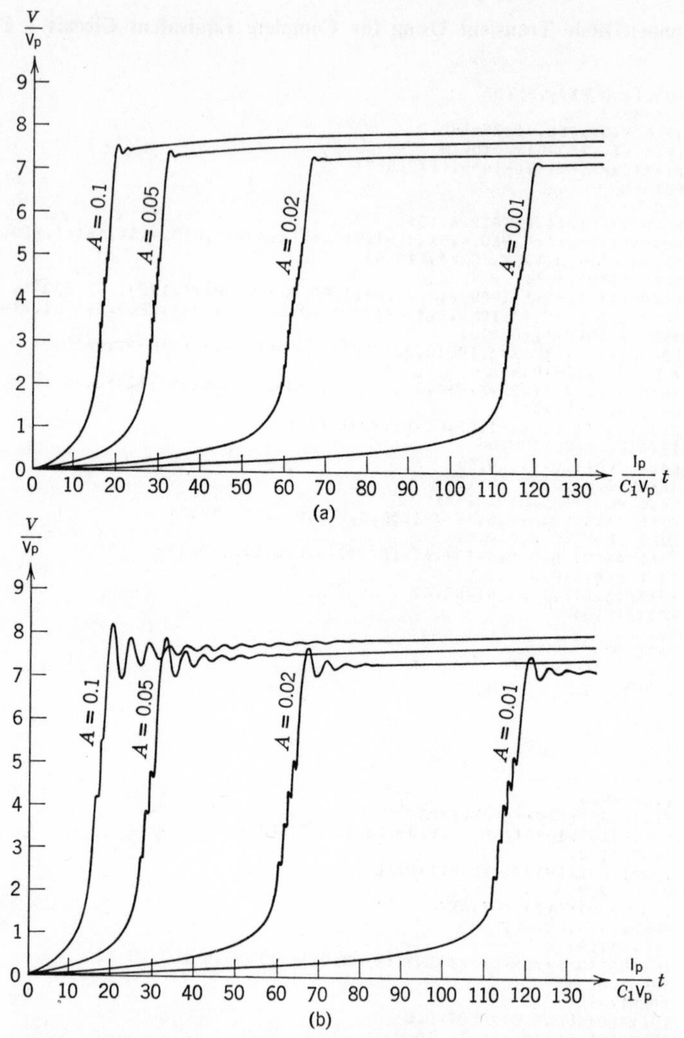

Figure 6.13 Transient response of the circuit of Fig. 6.10. Normalized voltage $V/\mathrm{V_p}$ is shown as a function of normalized time, $(\mathrm{I_p}/C_1\mathrm{V_p})t$, with normalized ramp-current slope, $A \equiv kC_1\mathrm{V_p}/\mathrm{I_p}^2$, as parameter.

(a) $C_2/C_1 = 0.05, \dfrac{\mathrm{I_p}^2 L}{\mathrm{V_p}^2 C_1} = 0.1;$

(b) $C_2/C_1 = 0.05, \dfrac{\mathrm{I_p}^2 L}{\mathrm{V_p}^2 C_1} = 0.5;$

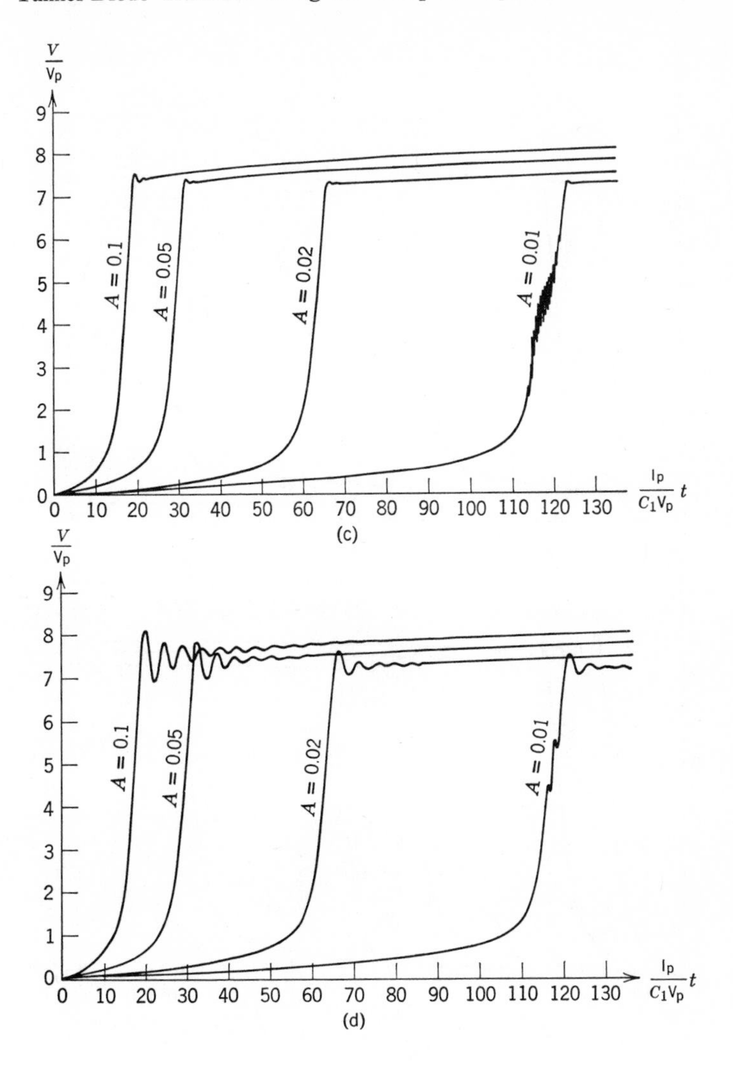

Figure 6.13 (Continued).

(c) $C_2/C_1 = 0.1, \dfrac{I_p{}^2L}{V_p{}^2C_1} = 0.1;$

(d) $C_2/C_1 = 0.1, \dfrac{I_p{}^2L}{V_p{}^2C_1} = 0.5;$

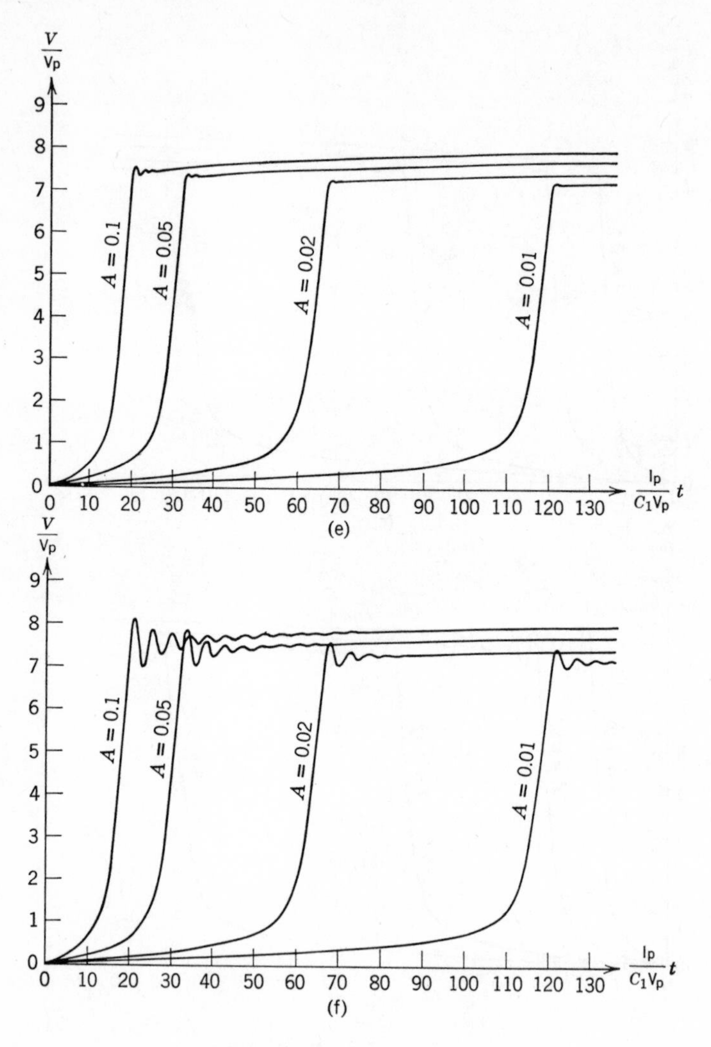

Figure 6.13 (Continued).

(e) $C_2/C_1 = 0.2, \dfrac{I_p^2L}{V_p^2C_1} = 0.1;$

(f) $C_2/C_1 = 0.2, \dfrac{I_p^2L}{V_p^2C_1} = 0.5.$

EXERCISES

1. In addition to the dc characteristics of Equation (6.1), a real tunnel-diode also has a series ohmic resistance. Sketch the *i-v* characteristic of a tunnel-diode with $V_p = 100$ mV, $I_p = 10$ mA, and a series ohmic resistance of 5 Ω.

2. Derive the incremental resistance $r_i \equiv (dI/dV)^{-1}$ from Equation (6.1). Plot $r_i I_p/V_p$ as function of V/V_p. At what voltage is the resistance minimum and what is its value, $r_{i_{\min}}$?

3. Use the results of Exercise 2 to calculate the time constants, $r_{i_{\min}}(C_1 + C_2)$ and $\sqrt{LC_2}$ of Fig. 6.2 for a tunnel-diode with V_p = 100 mV, I_p = 10 mA, $C_1 = C_2 = 1$ pF, and $L = 0.5$ nH.

4. Use Fig. 6.9 to calculate the 2 V_p to 7 V_p risetime for the circuit of Fig. 6.4. Assume $k = 10$ mA/ns, $C = 10$ pF, $I_p = 10$ mA, and $V_p = 100$ mV.

5. Replace the current-ramp of Exercise 4 with a current function with a linear 0% to 100% risetime of 2 ns (see Fig. 6.14). Use Fig. 6.7 to show that the resulting 2 V_p to 7 V_p risetime is the same as that of Exercise 4.

6. Compute the time required to reach $V = 200$ mV, when the current-ramp in Exercise 4 is replaced by a current function of 20 mA as in the preceding exercise, but with a linear 0% to 100% risetime of 0.02 ns, 0.2 ns, 2 ns, and 20 ns. Use other parameters from Exercise 4 above.

7. Find the 2 V_p to 7 V_p risetime for the circuit of Fig. 6.10 assuming $k = 5$ mA/ns, $C_1 = 10$ pF, $I_p = 10$ mA, $V_p = 100$ mV, and $L = 0.5$ nH. Carry out the computation for diodes with $C_2 = 0.5$ pF, 1 pF, and 2 pF. Which of these three diodes is preferable if the fastest 2 V_p to 7 V_p risetime is desired without ringing on the rising edge?

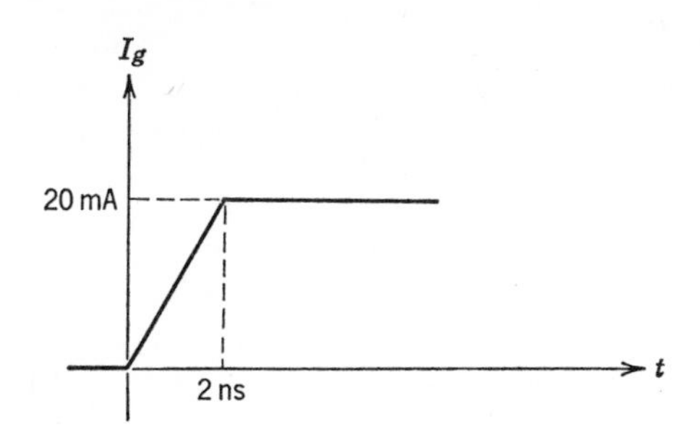

Figure 6.14

8. Show that in the circuit of Fig. 6.4 the voltage for $t > 0$ can be approximated as $V(t)$ = constant $\times\ t^2$ if $A \equiv kCV_p/I_p^2 \gg 1$. Derive the value of the constant. Use this approximation to compute $V(t)$ for $A = 10$ and compare the result with the transient shown in Fig. 6.7.

9. Sketch the transient in the circuit of Fig. 6.4 for $A = 0.01$, by neglecting capacitance C and compare it with the result of Fig. 6.7 which includes capacitance C.

10. Demonstrate that the period of the ringing on the rising edge in Fig. 6.13(b), (c), and (d) is in the vicinity of $2\pi\sqrt{LC_2}$.

ooooo

7

ooooo

Junction Transistors

ooooooooooooooooooooooooooooooo

A junction transistor (or bipolar transistor) contains two P-N junctions. As a result, many basic properties can be derived from those of junction diodes, and the following discussion will draw heavily on the material presented in Chapter 5.

THE PNP TRANSISTOR

A schematic representation of a PNP junction transistor is shown in Fig. 7.1. Terminal E is the emitter, terminal C is the collector, and terminal B is the base. Two semiconductor junctions are formed, one between the emitter and the base, and the other one between the base and the collector.

If the collector were open-circuited, I_E would be given by

$$I_E = I_{ES}(e^{V_{BE}/V_T} - 1), \tag{7.1}$$

where I_{ES} is the reverse saturation current of the emitter-base diode.

If two voltage sources are connected with polarities as shown in Fig. 7.1, the emitter-base diode is forward-biased, and the collector-base diode is reverse-biased; under these conditions, the transistor is

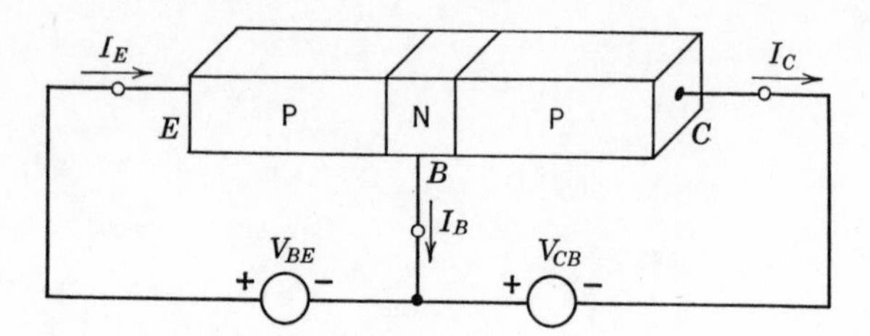

Figure 7.1 PNP junction transistor.

in the forward active region of operation. If, in addition, the base region is thin, most of I_E is drawn to the collector and only a small fraction will flow through the base terminal. The ratio of I_C and I_E is defined as α,

$$\alpha \equiv \frac{I_C}{I_E}, \qquad (\alpha < 1) \tag{7.2}$$

and the ratio of I_C and I_B as β:*

$$\beta \equiv \frac{I_C}{I_B}. \tag{7.3}$$

Since $I_E = I_B + I_C$, it follows that

$$\alpha = \frac{\beta}{1 + \beta} \tag{7.4}$$

and

$$\beta = \frac{\alpha}{1 - \alpha}. \tag{7.5}$$

For a typical transistor at normal operating levels, β may be between 10 and 1000, corresponding to an α of approximately 0.9 to 0.999.

A portion of the collector current results from the diode current of the reverse-biased collector-base diode; this current I_{CB} is given by the diode current with the emitter open-circuited:

$$I_{CB} = -I_{CS}(e^{-V_{CB}/V_T} - 1). \tag{7.6}$$

When $V_{CB} \gg V_T$, $I_{CB} \approx I_{CS}$. This current can be quite significant in circuits using transistors operating at low currents; it can, however, be neglected in many high-speed circuits, and will be done so in what follows. Also ignored will be variations of β as function of V_{CB} and

* β is often denoted by h_{FE}.

I_C. Under these assumptions, the dc behavior of the transistor in the forward active region is described by Equations (7.1), and (7.2) or (7.3), with α and β as constants of the transistor.

Another region of operation is the cut-off region where both junctions are reverse biased and only very small currents flow in the collector and in the emitter. A further region—finding widespread application in switching circuits of moderate speeds—is the saturated region of operation where both junctions are forward biased and the collector to emitter voltage may be very low. Finally, the emitter and collector voltages can be reversed from Fig. 7.1, resulting in the reverse active region of operation. Of these possibilities, the forward active and the cut-off regions are most frequently utilized in high-speed circuits, and all further discussion will be restricted to these two operating regions.

In addition to dc characteristics, inductances and capacitances of the transistor are also important in high-speed circuits. The inductances are primarily those of the connecting leads, and the capacitances are primarily those of the emitter-base and collector-base diodes. In the forward active region of operation, for the forward-biased emitter-base diode the diffusion capacitance of Equation (5.9) is usually dominant,* while for the reverse-biased collector-base diode the transition capacitance of Equation (5.12) and the stray capacitances dominate.

Based on the above discussion, the most prominent high-speed properties of a PNP transistor can be summarized by the equivalent circuit (model) of Fig. 7.2.** Resistor r_b (previously neglected) is the ohmic resistance of the semiconductor body of the base, a constant; current I_D is given by the diode equation

$$I_D = I_{ES}(e^{V_{B1}/V_T} - 1). \tag{7.7}$$

The base charge Q_B results in the diffusion capacitance C_e:

$$C_e \equiv \frac{dQ_B}{dV_{B1}} = \frac{dQ_B}{dI_D}\frac{dI_D}{dV_{B1}} = \frac{\tau_0}{r_e} \tag{7.8a}$$

* In many high-speed transistors, there are significant contributions by the transition capacitance and by stray capacitances. For more details see Ghosh et al. in references.

** Known as a hybrid-pi equivalent circuit.

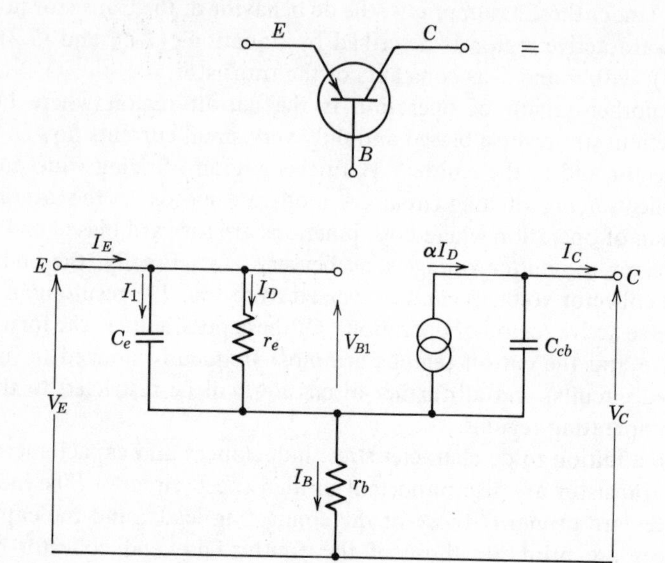

Figure 7.2 Hybrid-pi equivalent circuit of the PNP transistor.

where

$$r_e \equiv \frac{dV_{B1}}{dI_D} = \frac{\mathrm{V_T}}{I_D + I_{ES}},$$

hence

$$C_e = \frac{\tau_0 I_{ES}}{\mathrm{V_T}}\, e^{V_{B1}/\mathrm{V_T}} = \tau_0 \frac{I_D + I_{ES}}{\mathrm{V_T}}. \qquad (7.8b)$$

The gain-bandwidth product, f_T, a basic high-speed parameter of the transistor, is related to τ_0 by $\tau_0 = 1/2\pi f_T$. The collector-base capacitance, C_{cb}, can be, in many cases, approximated by a constant.

THE NPN TRANSISTOR

The operation of an NPN transistor is identical to that of a PNP transistor with all voltage and current polarities inverted (Fig. 7.3). A high-speed equivalent circuit of an NPN transistor is shown in

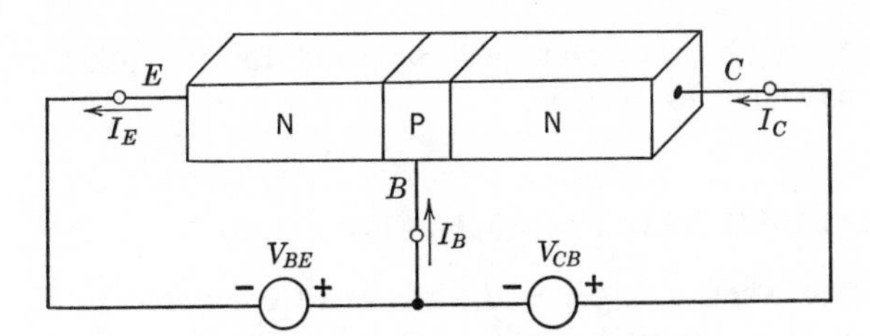

Figure 7.3 NPN junction transistor.

Fig. 7.4. Current I_D is determined by the diode equation

$$I_D = I_{ES}(e^{-V_{B1}/V_T} - 1), \tag{7.9}$$

capacitance C_e is the diffusion capacitance

$$C_e = \frac{\tau_0 I_{ES}}{V_T} e^{-V_{B1}/V_T} = \tau_0 \frac{I_D + I_{ES}}{V_T}, \tag{7.10}$$

and collector-base capacitance C_{cb} can be again approximated by a constant. In the following, circuits will be analyzed for only one

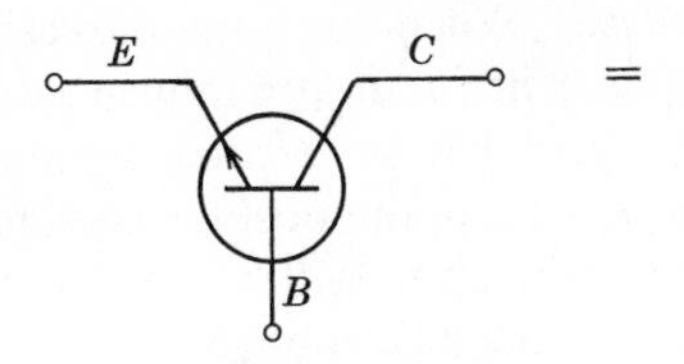

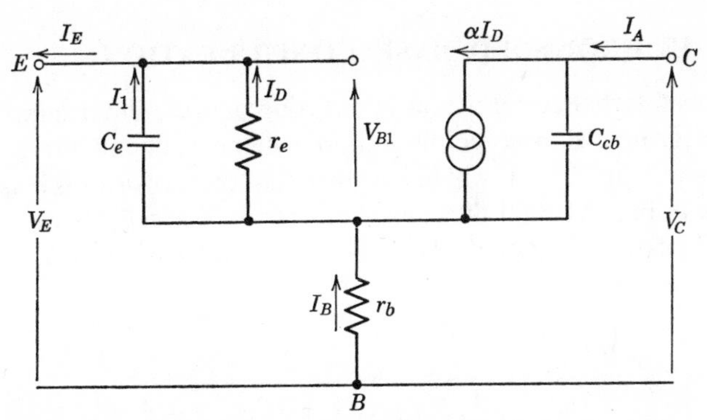

Figure 7.4 Hybrid-pi equivalent circuit of the NPN transistor.

type; the results, however, will be valid for the other type by appropriate changes of signs.

SMALL-SIGNAL PARAMETERS

In the case when changes in I_D are small compared to I_D itself, the I_D versus V_{B1} function in Fig. 7.2 or Fig. 7.4 [Equations (7.7) or (7.9)] may be replaced by a dc current (or voltage) source and an incremental resistance of

$$r_e = \frac{V_T}{I_D + I_{ES}} . \tag{7.11}$$

If this substitution is made, the resulting equivalent circuit is the so-called hybrid-pi *small-signal* equivalent circuit, which will be used quite often. The parameters are determined by the operating point, although the source establishing it is frequently not shown. When the signal currents are negligible in magnitude compared to the operating current, the small-signal equivalent circuit can be used directly for analysis. When this is not the case, the parameters of the small-signal equivalent circuit are influenced by the signal and have to be changed when changes become significant ("piecewise linear approximation"), or continuously (see Chapter 8). The small-signal nature of the analysis will be emphasized by the use of lower case letters for currents and voltages.

THE GROUNDED BASE CONFIGURATION

The small-signal current transfer function and the input impedance of the grounded base configuration will be analyzed for a PNP transistor (Fig. 7.5). It will be assumed that the transistor is biased as shown in Fig. 7.1, that the equivalent circuit of Fig. 7.2 is valid, and that C_{cb} can be neglected. With Equations (7.8) and (7.11),

$$\frac{\mathscr{L}\{i_D\}}{\mathscr{L}\{i_g\}} = \frac{\dfrac{1}{sC_e}}{\dfrac{1}{sC_e} + r_e} = \frac{1}{1 + sC_e r_e} = \frac{1}{1 + s\tau_0} \tag{7.12a}$$

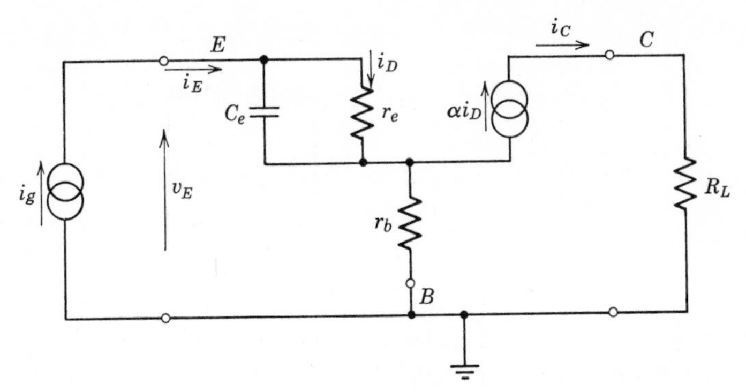

Figure 7.5 Grounded base configuration including small-signal PNP equivalent circuit.

and the current transfer function becomes

$$\frac{\mathscr{L}\{i_C\}}{\mathscr{L}\{i_g\}} = \frac{\alpha}{1 + s\tau_0}. \tag{7.12b}$$

The small-signal input impedance $\mathscr{L}\{v_E\}/\mathscr{L}\{i_E\}$ seen at the emitter can be shown to be as in Fig. 7.6. The input impedance can be made purely resistive by connecting the network of Fig. 7.7 in

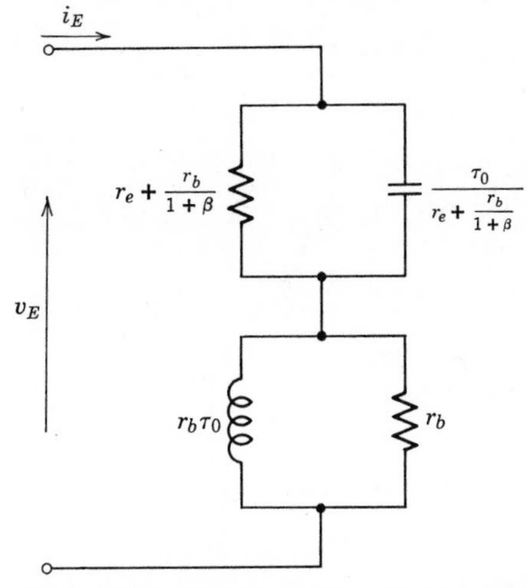

Figure 7.6 Small-signal input impedance of the grounded base configuration.

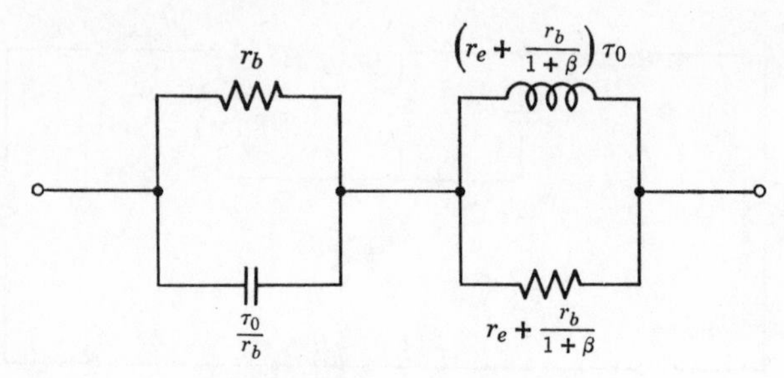

Figure 7.7 Series compensating network for the input of the grounded base configuration.

series with the emitter; the resulting overall input resistance is $r_e + r_b(\beta + 2)/(\beta + 1) \approx r_e + r_b$.

THE GROUNDED EMITTER CONFIGURATION

Making assumptions similar to those of the grounded base configuration, the grounded emitter configuration for an NPN transistor can be described as shown in Fig. 7.8. The small-signal current transfer function can be shown to be

$$\frac{\mathscr{L}\{i_C\}}{\mathscr{L}\{i_g\}} = \frac{\beta}{1 + s\tau_0(\beta + 1)}, \tag{7.13}$$

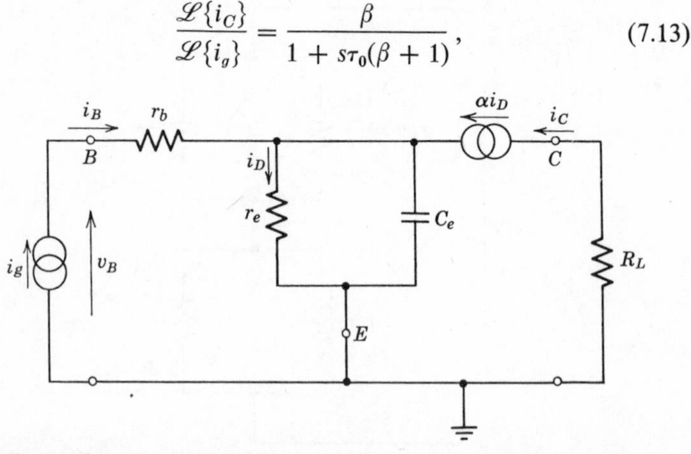

Figure 7.8 Grounded emitter configuration including small-signal NPN equivalent circuit.

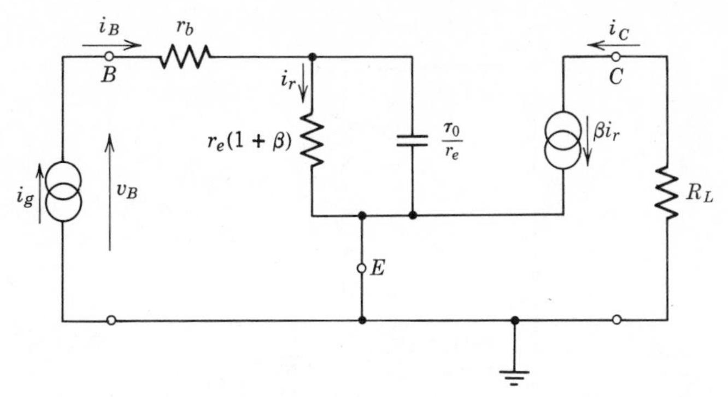

Figure 7.9 Alternate representation of the grounded emitter configuration of Fig. 7.8.

and the base input impedance can likewise be calculated as

$$\frac{\mathscr{L}\{v_B\}}{\mathscr{L}\{i_g\}} = r_b + \frac{r_e(1+\beta)}{1 + s\tau_0(1+\beta)}. \tag{7.14}$$

These relations suggest the new equivalent circuit shown in Fig. 7.9.

THE GROUNDED COLLECTOR CONFIGURATION (EMITTER FOLLOWER)

The grounded collector configuration, also known as emitter follower, is shown in Fig. 7.10. Using the same assumptions as

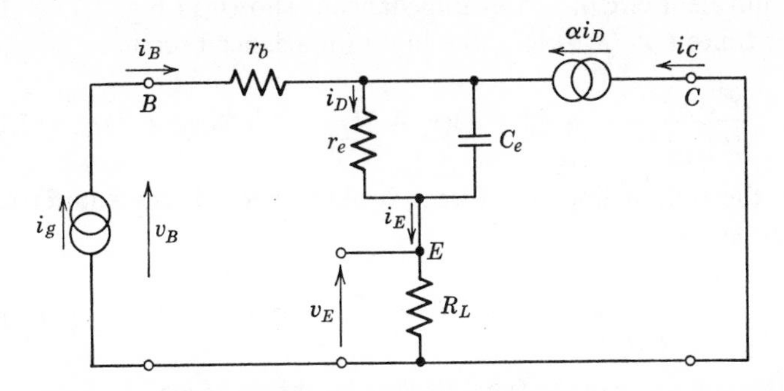

Figure 7.10 Grounded collector configuration including small-signal equivalent circuit.

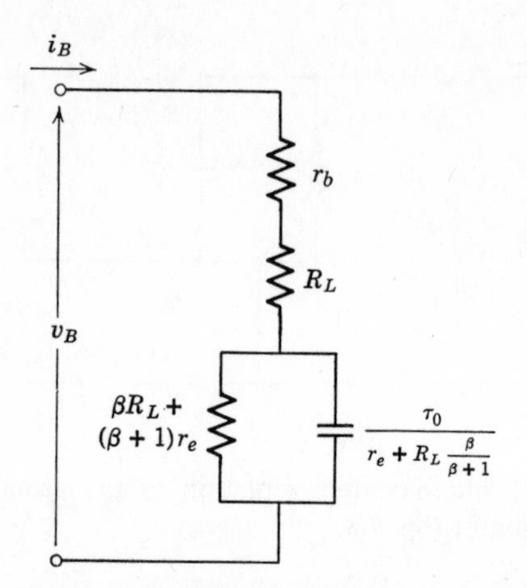

Figure 7.11 Small-signal input impedance of the grounded collector stage with the output loaded by a resistance R_L.

before, the current transfer function can be written as

$$\frac{\mathscr{L}\{i_E\}}{\mathscr{L}\{i_g\}} = 1 + \frac{\beta}{1 + s\tau_0(\beta + 1)}. \tag{7.15}$$

Also, the input impedance for a resistive load R_L can be shown to be

$$\frac{\mathscr{L}\{v_B\}}{\mathscr{L}\{i_B\}} = r_b + R_L + \frac{\beta R_L + (\beta + 1)r_e}{1 + s\tau_0(\beta + 1)}; \tag{7.16}$$

the equivalent circuit of this impedance is shown in Fig. 7.11.

For times $t \gg (\beta + 1)\tau_0$, the input impedance becomes

$$\frac{\mathscr{L}\{v_B\}}{\mathscr{L}\{i_B\}} = r_b + (\beta + 1)(r_e + R_L); \qquad t \gg (\beta + 1)\tau_0. \tag{7.17}$$

Also, the output impedance seen looking back at the emitter can be written as

$$\frac{\mathscr{L}\{-v_E\}}{\mathscr{L}\{i_E\}} = r_e + \frac{r_b}{\beta + 1}; \qquad t \gg (\beta + 1)\tau_0, \tag{7.18}$$

and the voltage transfer function can be shown to be

$$\frac{\mathscr{L}\{v_E\}}{\mathscr{L}\{v_B\}} = \frac{R_L}{R_L + r_e + r_b/(\beta + 1)}; \qquad t \gg (\beta + 1)\tau_0. \tag{7.19}$$

Thus, for times $t \gg (\beta + 1)\tau_0$, the emitter follower with a resistive load has a "high" ($>\beta R_L$) input impedance and a "low" output impedance. Also, for $R_L \gg r_e + r_b/(\beta + 1)$ the gain approaches 1; in this case the output voltage follows the input voltage: hence, the name emitter follower.

TRANSIENT RESPONSE OF THE EMITTER FOLLOWER

The small-signal transient response of the emitter follower will be analyzed for the more general case of the *R-C* load network shown in Fig. 7.12(a). The hybrid-pi equivalent circuit of Fig. 7.4 with

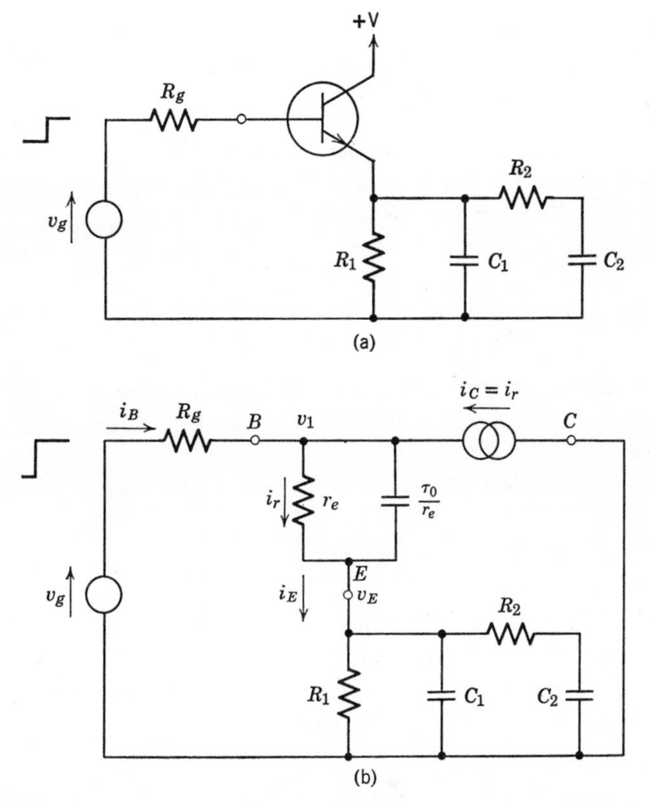

Figure 7.12 Emitter follower (a), and small signal equivalent circuit (b).

$\beta \to \infty$ is utilized in Fig. 7.12(b), where also the ohmic base resistance r_b is included in the generator resistance R_g and C_{cb} is again neglected. The following network equations can be written:

$$i_B = \frac{v_g - v_1}{R_g}, \tag{7.20}$$

$$i_E = i_B + i_C, \tag{7.21}$$

$$\mathscr{L}\{i_C\} = \frac{\mathscr{L}\{i_E\}}{1 + s\tau_0}, \tag{7.22}$$

$$\mathscr{L}\{i_E\} = \frac{\mathscr{L}\{v_1 - v_E\}(1 + s\tau_0)}{r_e}, \tag{7.23}$$

and

$$\mathscr{L}\{i_E\} = \frac{\mathscr{L}\{v_E\}}{Z_L} \tag{7.24}$$

where Z_L denotes the complex impedance of the load network (see Exercise 13, Chapter 2). For a unit step input voltage, $v_g = u(t)$, $\mathscr{L}\{v_g\} = 1/s$. With this, by utilizing Equations (7.20) through (7.24), $\mathscr{L}\{v_E\}$ becomes

$$\mathscr{L}\{v_E\} = \frac{1}{s}\frac{Z_L(1 + s\tau_0)}{R_g s\tau_0 + r_e + Z_L(1 + s\tau_0)}. \tag{7.25}$$

The general solution of Equation (7.25) will not be attempted, but the output voltage transient will be examined for three special cases of load impedance Z_L.

(a) *Pure resistive load:* $C_1 = 0$, $C_2 = 0$. In this case, $Z_L = R_1$ and

$$\mathscr{L}\{v_E\} = \frac{1}{s}\frac{R_1(1 + s\tau_0)}{r_e + R_1 + (R_g + R_1)s\tau_0}. \tag{7.26}$$

Hence, the roots of the denominator are always real; the output transient is free of ringing and of oscillation.

(b) *Resistive load with capacitances* $C_1 = \tau_0/R_1$ and $C_2 = 0$. In this case, the impedance of the load is

$$Z_L = \frac{R_1}{1 + s\tau_0},$$

and

$$\mathscr{L}\{v_E\} = \frac{R_1}{R_1 + r_e} \frac{1}{s} \frac{1}{1 + \dfrac{R_g}{R_1 + r_e} s\tau_0}. \tag{7.27}$$

Since again, as in case (a), the roots of the denominator are always real, there is no ringing or oscillation on the transient.

(c) *Large* R_1: $R_1 \gg R_g$, thus,

$$Z_L = \frac{1 + R_2 C_2 s}{s(C_1 + C_2 + R_2 C_1 C_2 s)}.$$

In order to analyze this case, it is convenient to introduce normalized parameters $C_1^* \equiv C_1 R_g/\tau_0$, $C_2^* \equiv C_2 R_g/\tau_0$, $G \equiv R_2/R_g$, and $E \equiv r_e/R_g$. Using these parameters, it can be shown that the denominator of Equation (7.25) becomes zero when

$$y(s) \equiv s^3 + p_1 s^2 + q_1 s + r_1 = 0 \tag{7.28}$$

where

$$p_1 \equiv \frac{C_1^* + C_2^*}{G C_1^* C_2^*} + E + \frac{1}{C_1^*}, \tag{7.29a}$$

$$q_1 \equiv \frac{C_1^* + C_2^*}{G C_1^* C_2^*} + \frac{1}{G C_1^* C_2^*} + \frac{1}{C_1^*}, \tag{7.29b}$$

and

$$r_1 \equiv \frac{1}{G C_1^* C_2^*}. \tag{7.29c}$$

The cubic equation (7.28) has three roots. If all the three roots are real, the transient response is free of oscillation and of ringing; if any of the three roots is complex, the transient response will have an oscillatory term of the form $e^{-at} \sin \omega t$.

To determine whether there are any complex roots, the following procedure will be followed: For each set of values of C_1^*, C_2^*, G, and E, it will be determined whether the function $y(s)$ of Equation (7.28) crosses the $y = 0$ axis three times; if it does, all three roots are real, if not, there is at least one complex root.*

The number of real zeros of $y(s)$ can be evaluated by an inspection of the sign of $y(s)$ at values of s where the derivative $dy/ds = 0$. From Equation (7.28),

$$\frac{dy}{ds} = s^2 + bs + c = 0, \tag{7.30}$$

where

$$b \equiv \tfrac{2}{3}p_1 \text{ and } c \equiv \tfrac{1}{3}q_1. \tag{7.31}$$

Furthermore, with

$$d \equiv \left(\frac{b}{2}\right)^2 - c = \left(\frac{p_1}{3}\right)^2 - \frac{q_1}{3}, \tag{7.32}$$

the roots of Equation (7.30) become

$$s_1 = -\frac{b}{2} - \sqrt{d} \tag{7.33a}$$

and

$$s_2 = -\frac{b}{2} + \sqrt{d}. \tag{7.33b}$$

The flow-chart of the procedure determining whether all roots of $y(s)$ are real for given parameters is shown in Fig. 7.13. Note that $y(s_1) = 0$ or $y(s_2) = 0$ are limiting cases where there are two real and equal roots at s_1 or s_2, respectively.

The results are summarized in Fig. 7.14. For $C_1R_g/\tau_0 \leq 0.2$ all three roots are real everywhere except between the two branches of the limits for a given r_e/R_g. For $C_1R_g/\tau_0 \geq 0.3$ all three roots are real between the two branches only.

* It can be shown that there are either two complex roots or none.

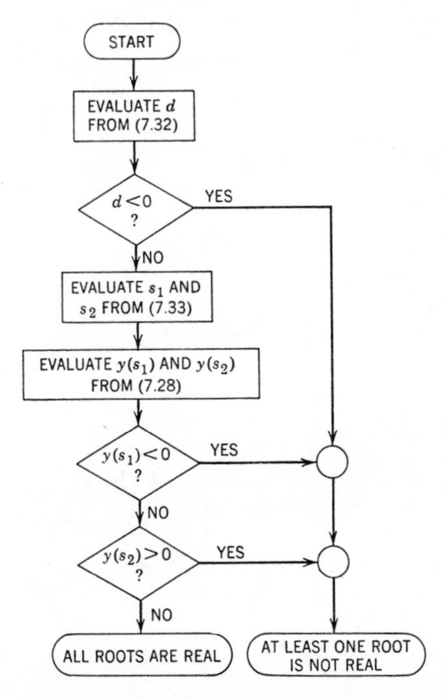

Figure 7.13 Flow-chart of the computer program to determine whether all three roots of Equation (7.28) are real.

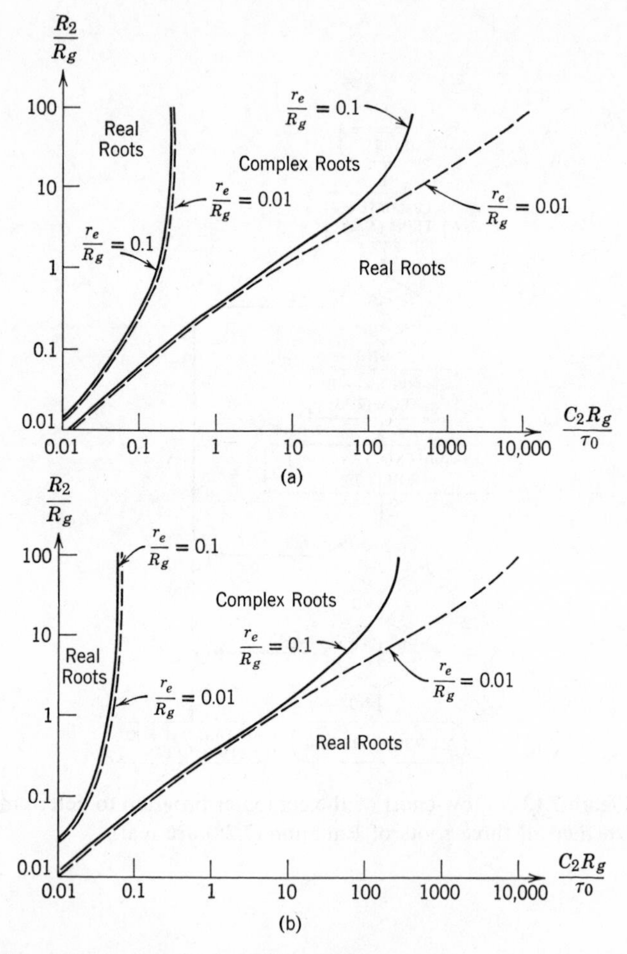

Figure 7.14 Roots of cubic equation (7.28) with $r_e/R_g = 0.1$ (solid lines) and $r_e/R_g = 0.01$ (broken lines).

(a) $C_1R_g/\tau_0 = 0.01$, (d) $C_1R_g/\tau_0 = 1$,
(b) $C_1R_g/\tau_0 = 0.2$, (e) $C_1R_g/\tau_0 = 10$,
(c) $C_1R_g/\tau_0 = 0.3$, (f) $C_1R_g/\tau_0 = 100$

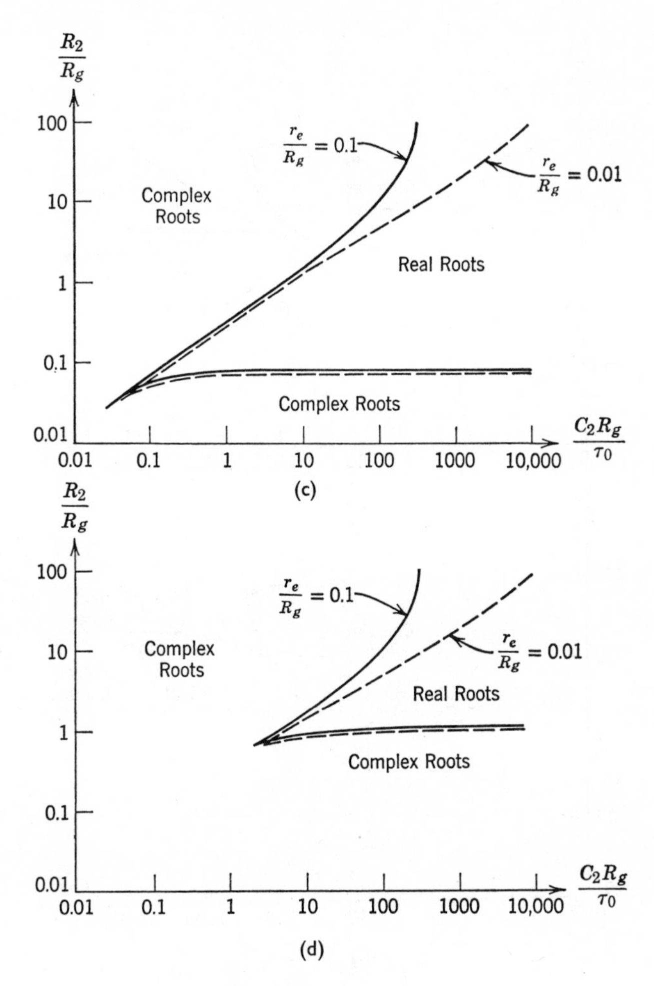

Figure 7.14 (Continued)

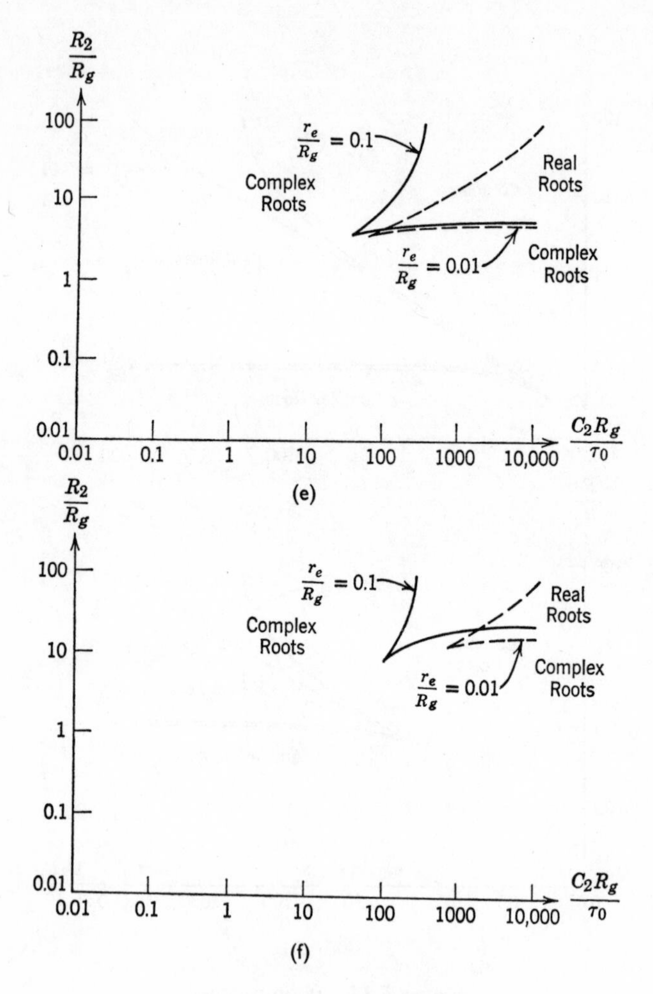

Figure 7.14 (Continued).

EXERCISES

1. Calculate the diffusion capacitance C_e and the incremental resistance r_e for a transistor with $f_T = 1$ GHz $= 10^9$ cycles per second, $V_T = 50$ mV, at an emitter current of 10 mA. Assume $I_{ES} \ll 10$ mA.

2. Calculate the small-signal input impedance of the circuit of Fig. 7.15 by utilizing Fig. 7.6. Assume $f_T = 1$ GHz $= 10^9$ cycles/second, $r_b = 20\ \Omega$, $\beta = 100$, $V_T = 50$ mV, and $I_{ES} \ll I_E$.

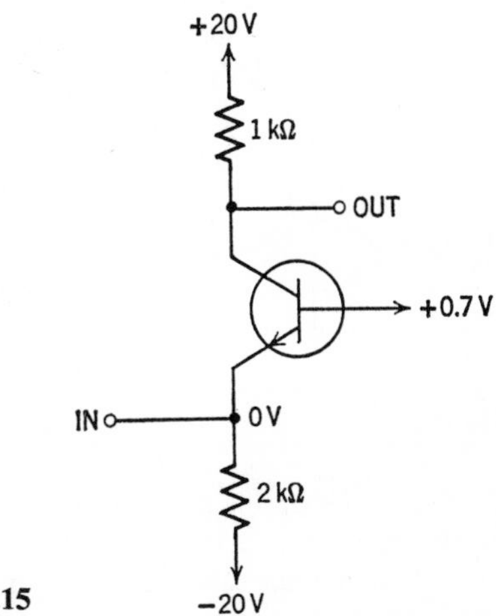

Figure 7.15

3. What is the approximate value of the 10% to 90% risetime of the collector current transient in the circuit of Exercise 2, if the input is driven by a current source of 0.1 mA $\times\ u(t)$.

4. Design a series compensating network that makes the input impedance of the circuit of Exercise 2 a pure resistance. What is the resulting input resistance? Note: $r_e + r_b \ll 2$ kΩ.

5. Show, that for a grounded emitter stage with $\beta \gg 1$, for $t \ll \beta\tau_0$, and for a unit step-function current input of $I_g = u(t)$, $I_C = tu(t)/\tau_0$.

6. Calculate the small-signal input impedance seen at the base of the emitter follower of Fig. 7.16 with the transistor parameters of Exercise 2 and with $C_{cb} = 0$.

7. Calculate the small-signal output impedance at the emitter of the emitter follower of Exercise 6. Repeat the calculation for the case when the base is driven from a generator with a source resistance of $R_g = 1000$ Ω.

8. Calculate, by utilizing Equation (7.26) the small-signal voltage transfer function of the emitter follower of Fig. 7.16 with the transistor parameters of Exercise 2 and with $C_{cb} = 0$.

9. Compute the small-signal transient response of the circuit of Fig. 7.16 with the transistor parameters of Exercise 2 and with $C_{cb} = 0$.

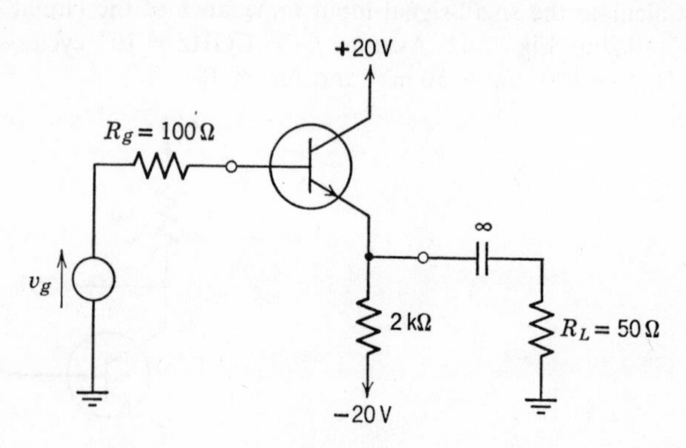

Figure 7.16

10. In the circuit of Fig. 7.17, $R_1 = 4000\ \Omega \gg R_g = 100\ \Omega$, hence, the results of Fig. 7.14 are applicable. Assume $\tau_0 = 0.1$ ns and estimate the minimum value of C_2 and the corresponding value of R_2 for a transient with no ringing for (a) $C_1 = 1$ pF and (b) $C_1 = 10$ pF. What is the allowable range of R_2 for a transient with no ringing if $C_1 = 1$ pF and $C_2 = 100$ pF?

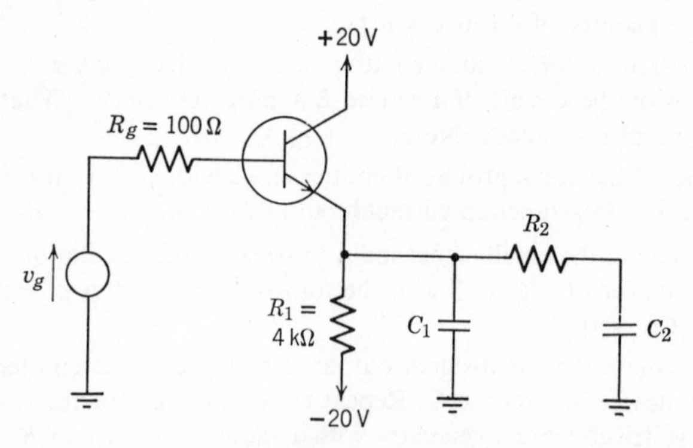

Figure 7.17

OOOOO

8

OOOOO

The Emitter-coupled Transistor Pair

OOOOOOOOOOOOOOOOOOOOOOOOOOOOOOOO

The emitter-coupled transistor pair of Fig. 8.1 has found many applications in high-speed switching circuits. There are several variations of the circuit such as: (a) Both bases may be driven, (b) one of the two collector resistors, R_{C1} or R_{C2}, may be omitted, (c) current source Q_3 may be replaced by a resistor.

In the following development it will be assumed that R_{C1} and R_{C2} are chosen such that both collector-to-base diodes are reverse biased; the circuit of Fig. 8.2 will be used as a reasonable approximation to the actual circuit. Transistors Q_1 and Q_2 are characterized by a single fixed parameter $\tau_0 \equiv 1/2\pi f_T$, where f_T is the gain-bandwidth product of each of the two transistors. Ohmic base resistances are included in R_g and all capacitances are lumped into C_{ext}. This approximation is reasonably good if R_{C1} is small so that the combination of the stray capacitance at the base of Q_1 and its collector-to-base capacitance can be indeed approximated by a single C_{ext}.

COMPUTATION OF THE SWITCHING TRANSIENT

The collector current $I_{C1}(t)$ will be computed for the generator voltage signal $V_g(t)$ of Fig. 8.3. The hybrid-pi equivalent circuit of

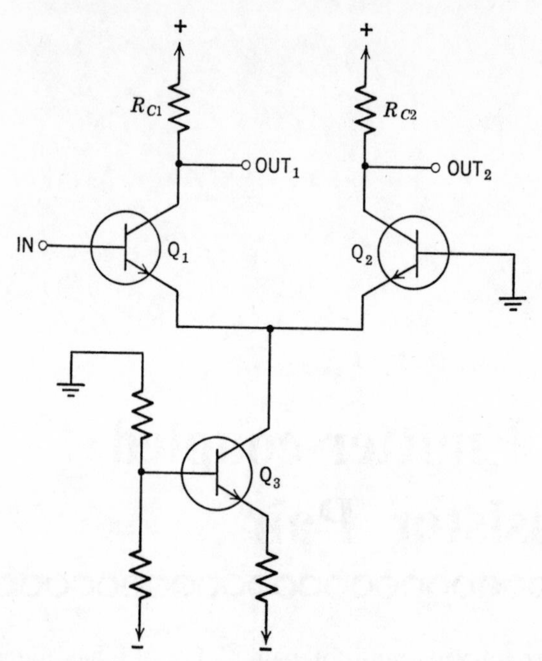

Figure 8.1 Basic circuit of the emitter-coupled transistor pair.

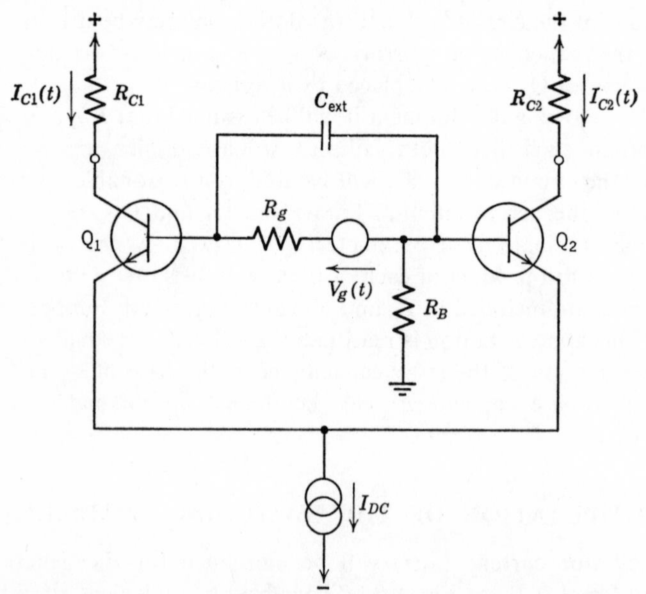

Figure 8.2 An approximation of the emitter-coupled transistor pair of Fig. 8.1.

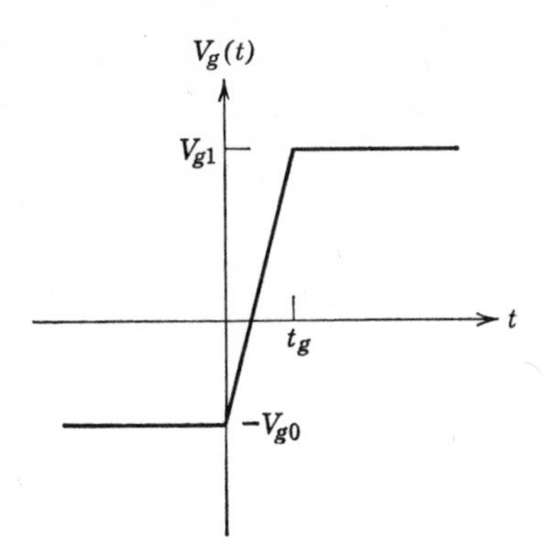

Figure 8.3 Generator voltage for Fig. 8.2.

Fig. 7.4 will be used for each transistor with $C_{cb} = 0$ and with $\alpha \approx 1$, i.e., $\beta \to \infty$. Under these assumptions the circuit in Fig. 8.4 results. It can be seen that the circuit enclosed in the box of broken lines is grounded only via R_B; hence, its value is arbitrary. It can also be shown that R_B can be omitted altogether, and will be so done in what follows. Also, by observing the nodes at B_1 and B_2, it is apparent that all of I_{B1} flows into C_{e1} and all of I_{B2} into C_{e2}. Thus, Fig. 8.4 can be redrawn as Fig. 8.5 where C_{ext} has been included in C'_{e1} and C'_{e2}.

Now the transient of the circuit can be computed solely from the loop of V_g, R_g, C'_{e1}, and C'_{e2}. By defining

$$V_{B1E} \equiv V_{B1} - V_E \tag{8.1}$$

and

$$V_{B2E} \equiv V_{B2} - V_E, \tag{8.2}$$

the collector currents are given by the diode equation as

$$I_{C1} = I_0(e^{V_{B1E}/V_T} - 1) \tag{8.3}$$

and

$$I_{C2} = I_0(e^{V_{B2E}/V_T} - 1); \tag{8.4}$$

also

$$I_{C1} + I_{C2} = I_{DC}. \tag{8.5}$$

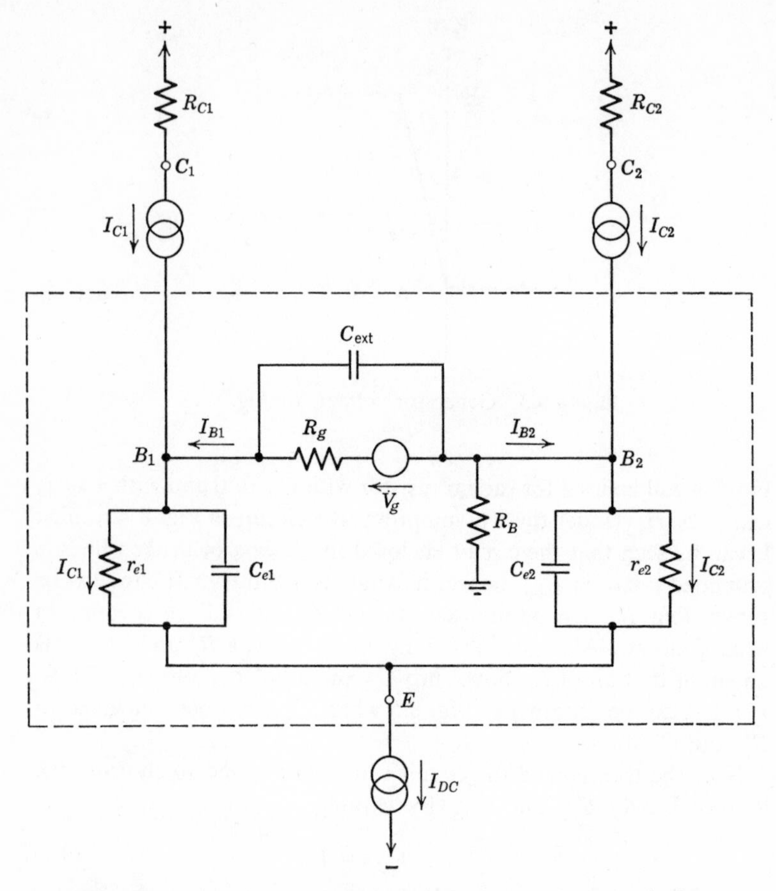

Figure 8.4 The circuit of Fig. 8.2 with the transistor equivalent circuit of Fig. 7.4.

If the charges in the base-emitter diodes are Q_{B1E} and Q_{B2E} then for identical transistors capacitances C_{e1} and C_{e2} are given by

$$C_{e1} \equiv \frac{dQ_{B1E}}{dV_{B1E}} = \frac{dI_{C1}\tau_0}{dV_{B1E}} = \frac{\tau_0}{\mathrm{V_T}} I_0 \mathrm{e}^{V_{B1E}/\mathrm{V_T}} \tag{8.6}$$

and

$$C_{e2} \equiv \frac{dQ_{B2E}}{dV_{B2E}} = \frac{dI_{C2}\tau_0}{dV_{B2E}} = \frac{\tau_0}{\mathrm{V_T}} I_0 \mathrm{e}^{V_{B2E}/\mathrm{V_T}}; \tag{8.7}$$

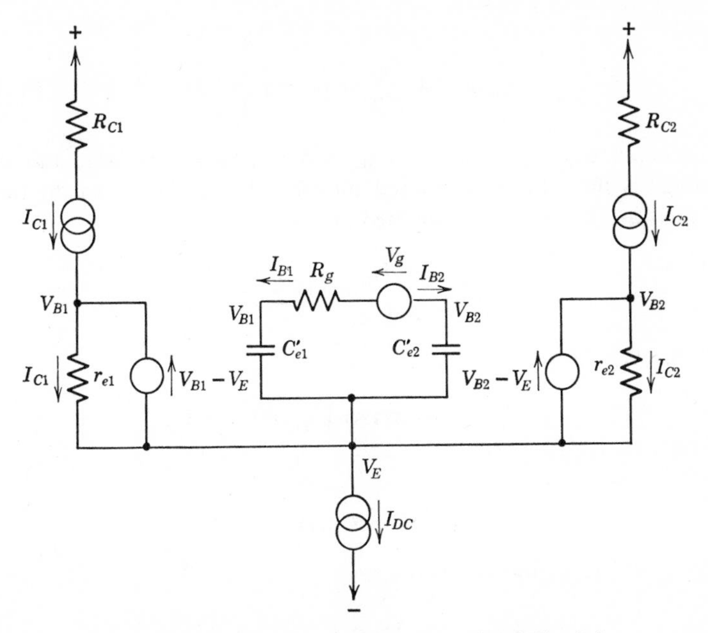

Figure 8.5 Simplification of the circuit of Fig. 8.4.

also, it can be shown that

$$C'_{e1} = C_{e1} + C_{\text{ext}} \frac{C_{e1} + C_{e2}}{C_{e2}}, \tag{8.8}$$

$$C'_{e2} = C_{e2} + C_{\text{ext}} \frac{C_{e1} + C_{e2}}{C_{e1}}, \tag{8.9}$$

and

$$I_{B1} = \frac{V_g + V_{B2E} - V_{B1E}}{R_g}. \tag{8.10}$$

The base-emitter voltages are given by the integrals

$$V_{B1E} = \int \frac{I_{B1}}{C'_{e1}}\, dt \tag{8.11}$$

and

$$V_{B2E} = \int \frac{I_{B2}}{C'_{e2}}\,dt = -\int \frac{I_{B1}}{C'_{e2}}\,dt. \tag{8.12}$$

Unfortunately, I_{B1}, r_{e1}, r_{e2}, C'_{e1}, and C'_{e2} vary with time and the integrals have to be evaluated numerically. In order to do this, Equation (8.11) is approximated as

$$V_{B1E} = \int \frac{I_{B1}}{C'_{e1}}\,dt \approx \sum \frac{I_{B1}}{C'_{e1}}\,\Delta t,$$

that can be also written as

$$V_{B1E}(t + \Delta t) \approx V_{B1E}(t) + \Delta V_1 \tag{8.13}$$

where

$$\Delta V_1 \equiv \frac{I_{B1}(t)}{C'_{e1}(t)}\,\Delta t. \tag{8.14}$$

Similarly, Equation (8.12) becomes

$$V_{B2E}(t + \Delta t) \approx V_{B2E}(t) + \Delta V_2 \tag{8.15}$$

where

$$\Delta V_2 \equiv \frac{-I_{B1}(t)}{C'_{e2}(t)}\,\Delta t. \tag{8.16}$$

The initial values of V_{B1E} and V_{B2E} can be computed from Equations (8.3), (8.4), and (8.5) as

$$V_{B1E}(t = 0) = V_g(t < 0) + \mathrm{V_T} \ln \frac{2 + I_{DC}/I_0}{1 + e^{V_g(t<0)/\mathrm{V_T}}}, \tag{8.17}$$

$$V_{B2E}(t = 0) = \mathrm{V_T} \ln \frac{2 + I_{DC}/I_0}{1 + e^{V_g(t<0)/\mathrm{V_T}}} \tag{8.18}$$

and the initial value of I_{C1} can be computed by substituting Equation (8.17) into Equation (8.3).

Equations (8.3), (8.8), (8.9), (8.10), (8.13), and (8.15) have been solved by using a digital computer and the programming flowchart of Fig. 8.6 with $\Delta t_{\max} = 0.01\,\tau_0$, $\Delta t_{\min} = 10^{-5}\,\tau_0$, and $\Delta V_{\max} = 0.01\,\mathrm{V_T}$. The Fortran-IV program is shown in Fig. 8.7.

RESULTS

Representative waveforms of $I_{C1}(t)$ are shown in Fig. 8.8. The rise-times between the 10% and 90% points of I_{C1}/I_{DC} (which is identical to that of I_{C2}/I_{DC}) are summarized in Table 8.1, together with those obtained from the following approximation:

$$t_r \approx \sqrt{(t_{r\tau} + t_{rc})^2 + t_{rg}^{\ 2}}, \tag{8.19}$$

where

$$t_{r\tau} \equiv 0.8 \frac{R_g I_{DC}}{V_{g1} - 0.4\mathrm{V_T}} \tau_0, \tag{8.20}$$

$$t_{rc} \equiv \frac{2\mathrm{V_T} \ln 9}{V_{g1} - 0.8\mathrm{V_T}} R_g C_{\text{ext}}, \tag{8.21}$$

and

$$t_{rg} \equiv \frac{2\mathrm{V_T} \ln 9}{V_{g1} + V_{g0}} t_g. \tag{8.22}$$

Thus, there are three contributions to the overall risetime: $t_{r\tau}$ of Equation (8.20) reflects the finite τ_0 of the transistors (finite-gain-bandwidth product), t_{rc} of Equation (8.21) results from the finite C_{ext}, and t_{rg} of Equation (8.22) originates from the finite risetime t_g of the input signal.

$\underline{t_{r\tau}}$. In the limiting case when C_{ext} and t_g are zero, the risetime is given by Equation (8.20). For $V_{g1} \gg \mathrm{V_T}$ this risetime is the current gain $R_g I_{DC}/V_{g1}$ multiplied by τ_0 and by a factor of 0.8 for a risetime measured between the 10% and 90% points. The term 0.4 $\mathrm{V_T} \approx 20$ mV in the denominator of Equation (8.20) represents a voltage "used up" for dc switching, which has to be taken into account if V_{g1} is not much greater than $\mathrm{V_T}$.

$\underline{t_{rc}}$. In addition to the charge $I_{DC}\tau_0$ in the base emitter junction, the source has to supply a charge to capacitor C_{ext} resulting in the risetime t_{rc} of Equation (8.21). The dc voltage swing on the bases between the 10% and 90% points of I_{C1} [from Equations (8.3), (8.4), and (8.5) with $I_{C1} \gg I_0$ and $I_{C2} \gg I_0$] is $2\mathrm{V_T} \ln 9$. Assuming a voltage of 0.8 $\mathrm{V_T}$ "lost" from V_{g1} for dc current transfer, t_{rc} is then the time required to supply the charge to C_{ext} during a voltage swing of $2\mathrm{V_T} \ln 9$.

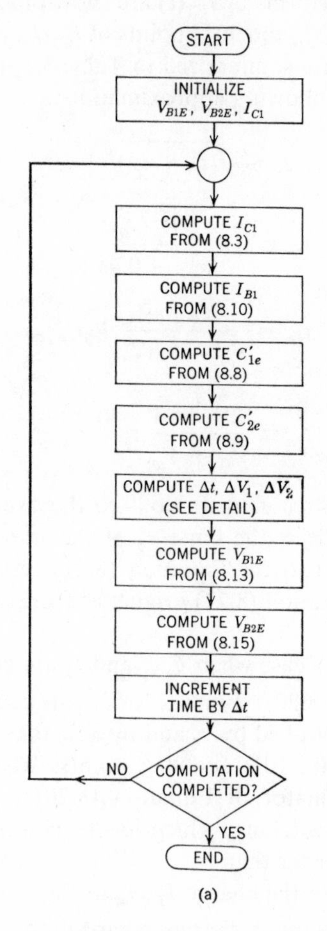

Figure 8.6 Flow-chart of the computer program.

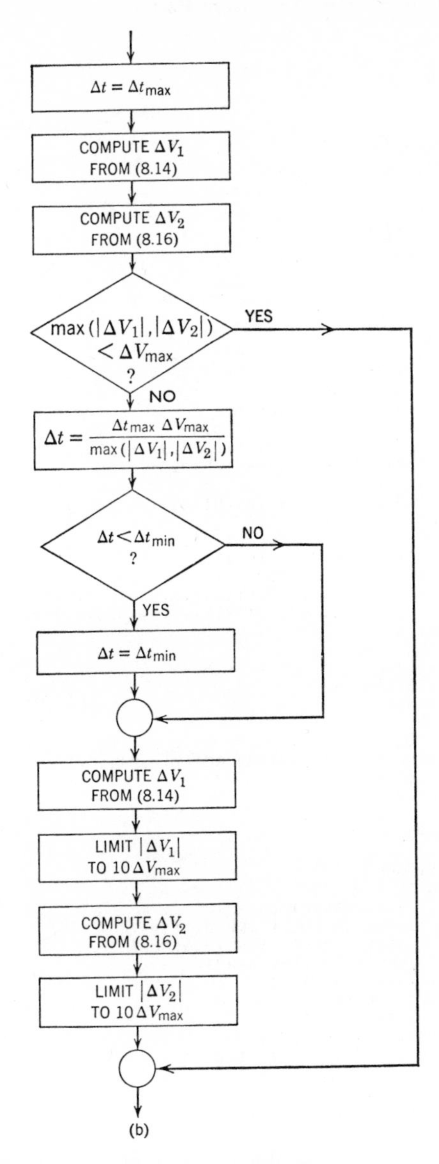

Figure 8.6 (Continued).

```
      FUNCTION CS(V)
      IF(V.LT.-100.0)V=-100.0
      IF(V.GT.+100.0)V=+100.0
      CS=EXP(V)
      RETURN
      END
      FUNCTION DIODE(V)
      IF(V.LT.-100.0)V=-100.0
      IF(V.GT.+100.0)V=+100.0
      DIODE=EXP(V)-1.0
      RETURN
      END
      REAL*8 I1,IC1,IZ,V1E,V2E,VE,VGZERO
      CALL STRTP1(10)
    1 FORMAT(5F10.5,I1,F9.5)
    2 FORMAT(' ',5F10.5,I1,F9.5)
    3 FORMAT(' ',1PE10.3,5(1PD10.3),4(1PE10.3))
    4 FORMAT(' ','RISETIME=',F10.4)
   99 CONTINUE
      READ(5,1)VGZERO,VG,R1,U1,DELT,NEWPLT,A
      IF(U1.LE.0)GO TO 100
      WRITE(6,2)VGZERO,VG,R1,U1,DELT,NEWPLT,A
      IF(NEWPLT.EQ.0)GO TO 11
      CALL PLOT1(20.0,0.0,-3)
      CALL AXIS1(0.0,0.0,'T',-1,8.0,0.0,0.0,10.,10.0)
      CALL AXIS1(0.0,0.0,'I',+1,5.0,90.0,0.0,0.2,10.0)
   11 CONTINUE
      VGZERO=-VGZERO
      TG=1.0/A
      IZ=EXP(-U1)
      IC1=0.0
      T1=0.0
      T2=0.0
      VE=DLOG((1.0+DEXP(VGZERO))/(2.0+1.0/IZ))
      V1E=VGZERO-VE
      V2E=-VE
      J=0
      XP=-0.02
      YPLOT=-0.1
   12 CONTINUE
      T=J*DELT
      I1=IZ*DIODE(V1E)
      IF(T1.EQ.0.0.AND.I1.GE.0.1)T1=T
      IF(T2.EQ.0.0.AND.I1.GE.0.9)T2=T
      V=VGZERO+A*T*(VG-VGZERO)
      IC1=(V-V1E+V2E)/R1
      C1=CS(V1E)*IZ
      DV1=IC1*DELT/C1
      IF(DV1.GT.1.0)DV1=1.0
      IF(DV1.LT.-1.0)DV1=-1.0
      C2=CS(V2E)*IZ
      DV2=IC1*DELT/C2
      IF(DV2.GT.1.0)DV2=1.0
```

Figure 8.7 Fortran-IV computer program for the generation of Fig. 8.8(a) and Fig. 8.8(b).

```
      IF(DV2.LT.-1.0)DV2=-1.0
      V1E=V1E+DV1
      V2E=V2E-DV2
      XPLOT=T*0.1
      IF(XPLOT-XP.LT.0.01)GO TO 14
      XP=XPLOT
      IF(I1.GE.0.998)GO TO 98
      YPL=5.0*I1
      IF(YPLOT.GE.YPL.AND.YPL.GE.4.95)GO TO 98
      YPLOT=YPL
      IF(YPLOT.LT.0.0)YPLOT=0.0
      IF(YPLOT.GE.10.0)YPLOT=10.0
      IF(J.NE.0)GO TO 13
      CALL PLOT1(XPLOT,YPLOT,+3)
   13 CONTINUE
      CALL PLOT1(XPLOT,YPLOT,+2)
   14 CONTINUE
      J=J+1
      IF(XPLOT.LE.8.0)GO TO 12
      TR=T2-T1
      WRITE(6,4)TR
      GO TO 99
   98 CONTINUE
      CALL PLOT1(XPLOT,5.0,+2)
      CALL PLOT1(8.0,5.0,+2)
      TR=T2-T1
      WRITE(6,4)TR
      GO TO 99
  100 CONTINUE
      CALL PLOT1(20.0,0.0,-3)
      CALL ENDP1
      STOP
      END
```

Figure 8.7 (Continued).

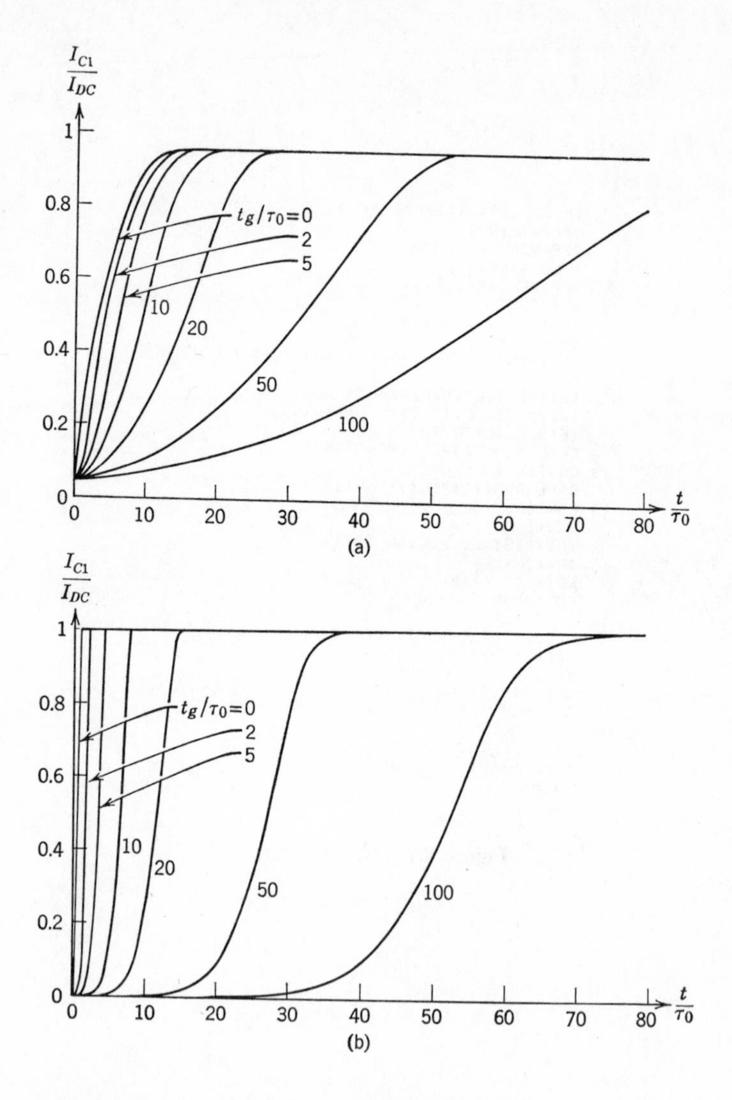

Figure 8.8 Representative waveforms of I_{C1}(t):
(a) $V_{g0}/V_T = 3$, $V_{g1}/V_T = 3$, $R_g I_{DC}/V_T = 30$, $C_{ext} = 0$
(b) $V_{g0}/V_T = 10$, $V_{g1}/V_T = 10$, $R_g I_{DC}/V_T = 30$, $C_{ext} = 0$
(c) $V_{g0}/V_T = 3$, $V_{g1}/V_T = 3$, $R_g I_{DC}/V_T = 30$, $t_g = 0$
(d) $V_{g0}/V_T = 10$, $V_{g1}/V_T = 30$, $R_g I_{DC}/V_T = 10$, $t_g = 0$

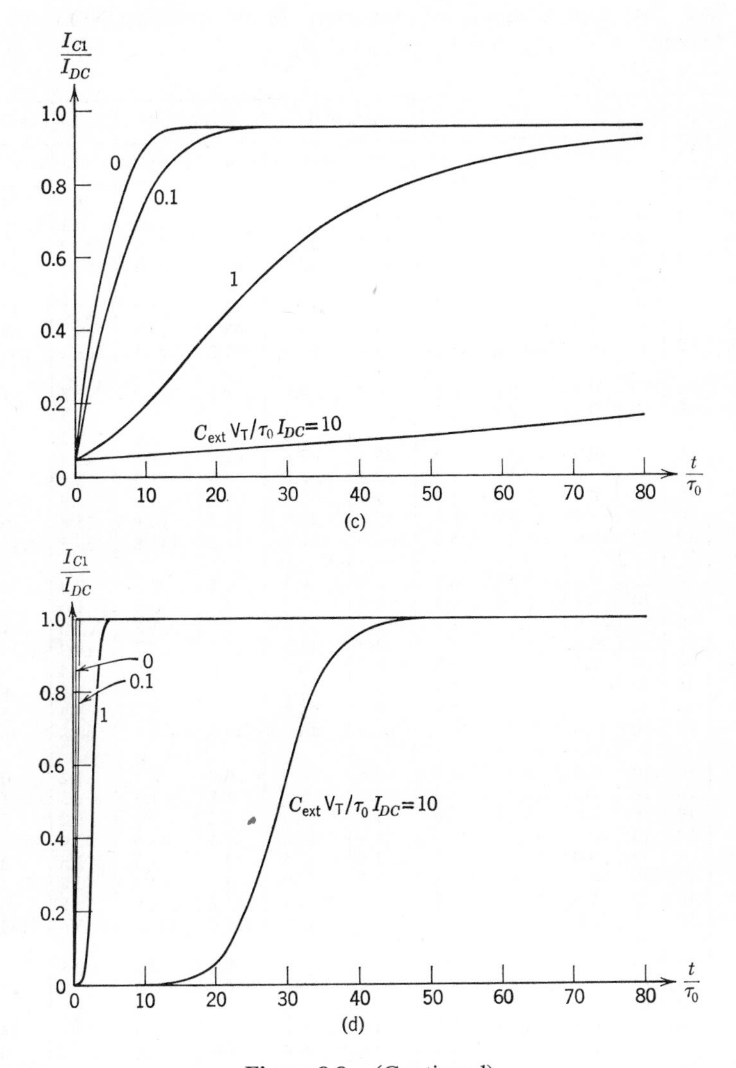

Figure 8.8 (Continued).

Table 8.1 10% to 90% risetime of i_{C1} for various values of V_{g0}, V_{g1}, R_g, C_{ext}, and t_g, using the flow-chart of Fig. 8.6 ("COMP"), and the approximation of Equation (8.19) ("APPROX"). Legend: $R \equiv R_g I_{DC}/\mathrm{V_T}$, $C \equiv C_{ext}\mathrm{V_T}/\tau_0 I_{DC}$.

$\frac{V_{g0}}{\mathrm{V_T}}$	$\frac{V_{g1}}{\mathrm{V_T}}$	R	C	$t_g/\tau_0 = 0$		$t_g/\tau_0 = 0.5$		$t_g/\tau_0 = 1$		$t_g/\tau_0 = 2$	
				COMP	APPROX	COMP	APPROX	COMP	APPROX	COMP	APPROX
3	3	10	0.0	3.1	3.1	3.2	3.1	3.3	3.2	3.7	3.4
3	3	10	0.1	5.0	5.1	5.0	5.1	5.0	5.1	5.3	5.3
3	3	10	1.0	21.7	23.1	21.7	23.1	21.7	23.1	21.7	23.1
3	3	10	10.0	189.7	202.8	189.7	202.8	189.7	202.8	189.7	202.8
3	3	30	0.0	9.5	9.2	9.5	9.2	9.5	9.3	9.7	9.3
3	3	30	0.1	15.0	15.2	15.0	15.2	15.1	15.2	15.1	15.3
3	3	30	1.0	65.4	69.2	65.4	69.2	65.4	69.2	65.4	69.2
3	3	30	10.0	561.3	608.5	561.3	608.5	561.3	608.5	561.3	608.5
3	10	10	0.0	0.8	0.8	0.9	0.9	1.1	0.9	1.5	1.1
3	10	10	0.1	1.2	1.3	1.3	1.3	1.4	1.4	1.7	1.5
3	10	10	1.0	5.2	5.6	5.2	5.6	5.2	5.6	5.3	5.7
3	10	10	10.0	45.4	48.6	45.4	48.6	45.4	48.6	45.4	48.6
3	10	30	0.0	2.4	2.5	2.4	2.5	2.5	2.5	2.9	2.6
3	10	30	0.1	3.7	3.9	3.7	3.9	3.8	3.9	3.9	4.0
3	10	30	1.0	15.8	16.8	15.8	16.8	15.8	16.8	15.8	16.8
3	10	30	10.0	136.3	145.8	136.3	145.8	136.3	145.8	136.3	145.8
3	30	10	0.0	0.3	0.3	0.4	0.3	0.6	0.3	0.9	0.4
3	30	10	0.1	0.4	0.4	0.5	0.4	0.7	0.4	1.0	0.5
3	30	10	1.0	1.7	1.8	1.7	1.8	1.7	1.8	2.0	1.8
3	30	10	10.0	14.9	15.3	14.9	15.3	14.9	15.3	14.9	15.3
3	30	30	0.0	0.8	0.8	0.9	0.8	1.0	0.8	1.4	0.9
3	30	30	0.1	1.2	1.3	1.3	1.3	1.4	1.3	1.7	1.3
3	30	30	1.0	5.2	5.3	5.2	5.3	5.2	5.3	5.2	5.3
3	30	30	10.0	44.8	46.0	44.8	46.0	44.8	46.0	44.8	46.0
10	10	10	0.0	0.8	0.8	0.8	0.8	0.9	0.9	1.2	0.9
10	10	10	0.1	1.2	1.3	1.2	1.3	1.2	1.3	1.3	1.4
10	10	10	1.0	5.2	5.6	5.2	5.6	5.2	5.6	5.2	5.6
10	10	10	10.0	45.4	48.6	45.4	48.6	45.4	48.6	45.4	48.6
10	10	30	0.0	2.4	2.5	2.4	2.5	2.4	2.5	2.6	2.5
10	10	30	0.1	3.7	3.9	3.7	3.9	3.7	3.9	3.7	4.0
10	10	30	1.0	15.8	16.8	15.8	16.8	15.8	16.8	15.8	16.8
10	10	30	10.0	135.7	145.8	135.6	145.8	135.7	145.8	135.6	145.8
10	30	10	0.0	0.3	0.3	0.4	0.3	0.5	0.3	0.8	0.3
10	30	10	0.1	0.4	0.4	0.4	0.4	0.5	0.4	0.7	0.5
10	30	10	1.0	1.7	1.8	1.7	1.8	1.7	1.8	1.7	1.8
10	30	10	10.0	14.9	15.3	14.9	15.3	14.9	15.3	14.9	15.3
10	30	30	0.0	0.8	0.8	0.8	0.8	0.9	0.8	1.2	0.8
10	30	30	0.1	1.2	1.3	1.2	1.3	1.2	1.3	1.2	1.3
10	30	30	1.0	5.2	5.3	5.2	5.3	5.2	5.3	5.2	5.3
10	30	30	10.0	44.8	46.0	44.8	46.0	44.8	46.0	44.8	46.0

$t_g/\tau_0 = 5$		$t_g/\tau_0 = 10$		$t_g/\tau_0 = 20$		$t_g/\tau_0 = 50$		$t_g/\tau_0 = 100$	
COMP	APPROX	COMP	APPROX	COMP	APPROX	COMP	APPROX	COMP	APPROX
5.2	4.8	8.2	7.9	15.1	15.0	36.8	36.7	73.3	73.3
6.4	6.3	9.0	8.9	15.5	15.5	37.0	37.0	73.4	73.4
21.8	23.3	22.8	24.2	26.2	27.3	41.6	43.3	75.2	76.8
189.7	202.9	189.7	203.0	189.7	203.4	191.1	206.1	201.5	215.6
10.7	9.9	13.0	11.8	18.5	17.3	38.5	37.8	74.0	73.8
15.7	15.7	17.3	16.9	21.8	21.1	39.8	39.7	74.9	74.8
65.4	69.3	65.4	69.5	66.5	70.7	75.0	78.3	97.0	100.7
561.2	608.5	561.2	608.5	561.1	608.6	560.7	609.6	560.3	612.9
2.7	1.9	4.5	3.5	7.7	6.8	17.3	16.9	34.0	33.8
3.0	2.1	4.8	3.6	8.1	6.9	17.7	17.0	34.2	33.8
5.8	5.9	7.4	6.5	11.3	8.8	21.6	17.8	37.5	34.3
45.4	48.6	45.4	48.7	45.8	49.1	51.9	51.5	68.5	59.2
4.1	3.0	6.3	4.2	10.0	7.2	19.9	17.1	35.8	33.9
4.9	4.3	6.9	5.2	10.8	7.8	21.0	17.4	37.1	34.0
15.8	16.9	16.4	17.2	19.0	18.1	30.3	23.9	48.2	37.8
136.3	145.8	136.3	145.8	136.3	146.0	136.7	146.8	143.8	149.7
1.5	0.7	2.3	1.4	3.7	2.7	7.6	6.7	13.9	13.3
1.7	0.8	2.6	1.4	4.0	2.7	8.0	6.7	14.3	13.3
2.9	1.9	4.3	2.2	6.4	3.2	11.2	6.9	18.1	13.4
14.9	15.3	15.3	15.4	17.5	15.5	26.7	16.7	39.4	20.3
2.4	1.0	3.5	1.6	5.4	2.8	9.9	6.7	16.5	13.3
2.7	1.4	4.0	1.8	6.0	2.9	10.7	6.8	17.5	13.4
5.7	5.4	7.1	5.5	10.2	6.0	17.2	8.5	26.1	14.3
44.8	46.0	44.8	46.0	44.9	46.0	50.0	46.4	64.4	47.8
2.0	1.4	3.2	2.3	5.4	4.5	11.7	11.0	22.3	22.0
1.9	1.7	3.2	2.6	5.5	4.6	12.0	11.1	22.6	22.0
5.2	5.7	5.2	6.0	6.0	7.1	12.3	12.3	23.2	22.7
45.4	48.6	45.4	48.6	45.4	48.8	45.4	49.8	45.4	53.3
3.2	2.7	4.7	3.3	7.3	5.1	14.2	11.3	24.6	22.1
3.8	4.1	4.5	4.5	7.1	5.9	14.4	11.7	25.3	22.3
15.8	16.9	15.8	17.0	15.8	17.4	16.8	20.1	26.3	27.7
135.6	145.8	135.5	145.8	135.4	145.9	135.1	146.2	134.5	147.4
1.3	0.6	2.0	1.1	3.2	2.2	6.5	5.5	11.7	11.0
1.2	0.7	1.9	1.2	3.2	2.2	6.6	5.5	12.0	11.0
1.7	1.9	2.3	2.1	3.5	2.8	6.9	5.8	12.3	11.1
14.9	15.3	14.9	15.4	14.9	15.5	15.0	16.3	20.3	18.9
2.0	1.0	3.0	1.4	4.7	2.3	8.6	5.6	14.2	11.0
1.8	1.4	2.7	1.7	4.3	2.5	8.4	5.6	14.4	11.1
5.2	5.4	5.2	5.4	5.5	5.8	9.5	7.7	15.4	12.2
44.8	46.0	44.8	46.0	44.8	46.0	44.8	46.3	44.8	47.3

t_{rg}. Equation (8.22) represents the risetime originating from that of the input signal during the voltage swing of $2V_T \ln 9$. Since t_{rg} is independent of the risetime of the circuit, $t_{rr} + t_{rc}$, the squares of the two risetimes are added in Equation (8.19).

As can be seen in Table 8.1, for $V_{g0} \geq 3\ V_T$ and $V_{g1} \geq 3\ V_T$, Equation (8.19) provides a reasonably good approximation and can be utilized to obtain risetimes for parameter values not listed in the Table.

EXERCISES

1. Show that voltages V_{C1} and V_{C2} in Fig. 8.9(a) are identical to V_{C1} and V_{C2} in Fig. 8.9(b) if C'_{e1} and C'_{e2} are given by Equations (8.8) and (8.9).

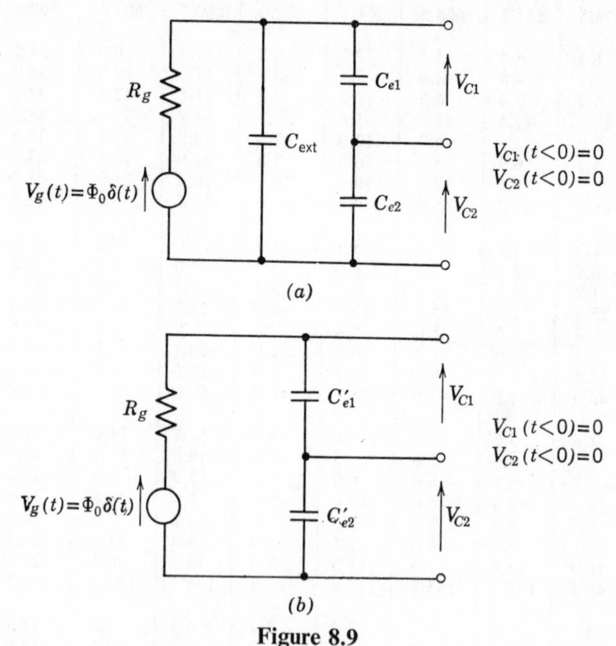

Figure 8.9

2. Derive Equations (8.17) and (8.18).

3. What fractional error in t_{r_r} results, if the second term in the denominator of Equation (8.20) is omitted? Plot this fractional error in percents versus V_{g1}/V_T for $1 \leq V_{g1}/V_T \leq 10$. Repeat above for t_{rc} of Equation (8.21).

4. Compare the risetime given by Equation (8.20) with the results of Exercise 5 in Chapter 7. Assume $V_{g1} \gg \mathrm{V_T}$.

5. Utilize Equations (8.3), (8.4), and (8.5) to show that if in Fig. 8.4 $I_{C1} \gg I_0$ and $I_{C2} \gg I_0$, then the dc voltage between the two bases B_1 and B_2 must change by $2\mathrm{V_T} \ln 9$ in order to change I_{C1} from 0.1 I_{DC} to 0.9 I_{DC}.

6. Calculate the 10% to 90% risetime t_r of Equation (8.19) for the circuit of Fig. 8.4. Assume $\mathrm{V_T} = 40$ mV, $f_T = 400$ MHz, $I_{DC} = 10$ mA, $C_{\mathrm{ext}} = 0$, $R_g = 120\ \Omega$, and the $V_g(t)$ of Fig. 8.3 with $V_{g0} = V_{g1} = 0.4$ V. Use $t_g = 0$ and $t_g = 2$ ns.

7. Repeat the computations of Exercise 6 with $C_{\mathrm{ext}} = 5$ pF and 10 pF.

8. Utilize Fig. 8.8 to compute the delay of the 50% point of I_{C1}/I_{DC} for the circuit of Exercise 6.

9. Assume that the risetime of Equation (8.19) is dominated by t_{rT} of Equation (8.20) and provide a reasoning for the absence of V_{g0} in t_{rT}.

10. Assume that the risetime of Equation (8.19) is dominated by t_{rc} of Equation (8.21) and provide a reasoning for the absence of V_{g0} in t_{rc}. In what way does V_{g0} influence the transient?

ooooo

9

ooooo

Digital Fanout Circuits

In high-speed digital circuitry with standard signal and impedance levels, a fanout circuit is frequently necessary in order to drive several loads from a source capable of driving one load only. The emitter-coupled transistor pair of Chapter 8 proves to be very suitable for this purpose. In this chapter, the risetimes of digital fanout circuits consisting of single and multiple stages of emitter-coupled pairs will be analyzed under the same assumptions and with the same equivalent circuit as in Chapter 8.

SINGLE-STAGE CIRCUIT

A single-stage emitter-coupled digital fanout circuit with a step-function driving source is shown in Fig. 9.1. Assuming identical transistors and $V_G \gg \mathrm{V_T}$, the signal risetime at the collector of $T1B$ can be approximated, by utilizing Equations (8.19), (8.20), and (8.21), as

$$t_r = \frac{1.6(R_g + 2r_b)(I_{DC}\tau_0 + C\Delta V)}{V_G} \tag{9.1}$$

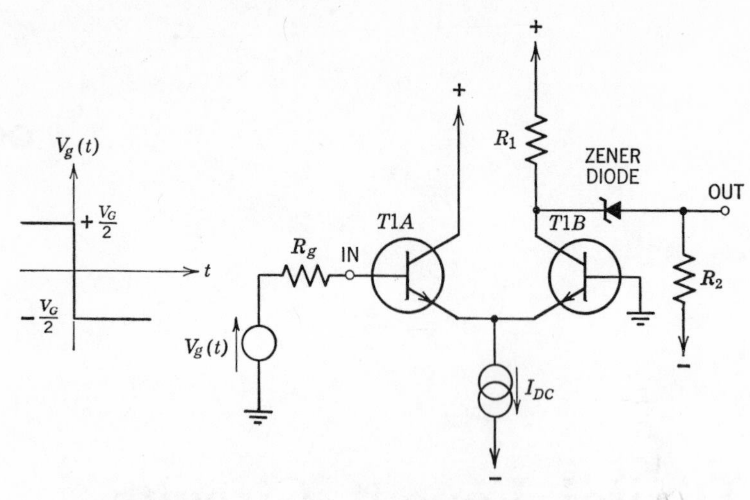

Figure 9.1 Single-stage digital fanout circuit.

where r_b is the ohmic base resistance, C is the collector-to-base capacitance, and

$$\Delta V \equiv \frac{2V_T \ln 9}{0.8}. \tag{9.2}$$

Defining the fanout, or current gain, as $F \equiv I_{DC}R_g/V_G$, t_r can be written as

$$t_r = 1.6\tau_0 F\left(1 + \frac{C\,\Delta V}{\tau_0 I_{DC}}\right)\left(1 + \frac{2r_b}{R_g}\right) = T_0 C^* R^*, \tag{9.3}$$

where

$$T_0 \equiv 1.6\tau_0 F, \tag{9.4a}$$

$$C^* \equiv 1 + \frac{C\,\Delta V}{\tau_0 I_{DC}}, \tag{9.4b}$$

and

$$R^* \equiv 1 + \frac{2r_b}{R_g}. \tag{9.4c}$$

The first factor, T_0, represents the risetime resulting from the finite gain-bandwidth product of the transistor, while C^* and R^* represent risetime multipliers introduced by C and r_b, respectively.

TWO-STAGE CIRCUIT

When the risetime of a single-stage fanout circuit is inadequate, the use of two stages should be considered. By choosing R_1 and R_2 in the first stage such that its output signal is symmetric with respect to the ground level, and connecting its output to the input of a similar stage having a higher I_{DC}, both stages will contribute current gain and risetime. In what follows, the parameters of the first and second stage will be identified by subscripts 1 and 2, respectively. The risetime of the collector current of $T1B$ in the first stage with a current gain of $F_1 \equiv I_{DC1}R_g/V_G$ can be written as

$$t_{r_1} = 1.6\tau_{0_1}F_1\left(1 + \frac{C_1\,\Delta V}{\tau_{0_1}I_{DC_1}}\right)\left(1 + \frac{2r_{b_1}}{R_g}\right). \tag{9.5}$$

If the charge supplied to the collector-to-base capacitance of $T1B$ in the first stage is negligible during the transient, if r_{b2} is small compared to R_1 and R_2, and if the voltage swing on the output of the first stage is much greater than V_T, then the second stage with a current gain of $F_2 \equiv I_{DC_2}/I_{DC_1}$ has an inherent risetime of

$$t_{r_2} = 1.6\tau_{0_2}F_2\left(1 + \frac{C_2\,\Delta V}{\tau_{0_2}I_{DC_2}}\right). \tag{9.6}$$

The resulting risetime of the two-stage circuit is thus [see Equation (8.19)]

$$t_r = (t_{r_1}^{\ 2} + t_{r_2}^{\ 2})^{1/2} = 1.6\tau_{0_1}\left\{\left[\frac{I_{DC_1}R_g}{V_G}\left(1 + \frac{C_1\,\Delta V}{\tau_{0_1}I_{DC_1}}\right)\left(1 + \frac{2r_{b_1}}{R_g}\right)\right]^2 + \left[\frac{\tau_{0_2}}{\tau_{0_1}}\frac{I_{DC_2}}{I_{DC_1}}\left(1 + \frac{C_2\,\Delta V}{\tau_{0_2}I_{DC_2}}\right)\right]^2\right\}^{1/2}. \tag{9.7}$$

Equation (9.7) shows that, with all other parameters constant, the risetime will be large both for small and for large values of I_{DC_1}. The optimum value of I_{DC_1} resulting in minimum risetime can be found from

$$\frac{dt_r^{\ 2}}{dI_{DC_1}} = 0. \tag{9.8}$$

This results in a fourth-order equation for F_1:

$$F_1^4 + \frac{C_1\,\Delta V}{\tau_{0_1} V_G/R_g} F_1^3 - \left(\frac{F_1 F_2 \dfrac{\tau_{0_2}}{\tau_{0_1}} + \dfrac{C_2\,\Delta V}{\tau_{0_1} V_G/R_g}}{1 + \dfrac{2r_{b_1}}{R_g}} \right)^2 = 0. \tag{9.9}$$

By introducing

$$C'^* \equiv \frac{C_1\,\Delta V}{\tau_{0_1} I_{DC_2}} \tag{9.10a}$$

and

$$K \equiv \frac{\dfrac{\tau_{0_2}}{\tau_{0_1}} + \dfrac{C_2\,\Delta V}{\tau_{0_1} I_{DC_2}}}{1 + \dfrac{2r_{b_1}}{R_g}}, \tag{9.10b}$$

Equation (9.9) becomes

$$F_1^2(1 + C'^* F_2) - (KF_2)^2 = 0, \tag{9.11}$$

from which the optimum current gain of the first stage, F_1, for a *given second stage current gain*, F_2, can be obtained as

$$F_1 = \frac{KF_2}{\sqrt{1 + C'^* F_2}}. \tag{9.12}$$

For identical transistors with negligible C and r_b, $C^* = 0$ and $K = 1$; hence, an $F_1 = F_2$, i.e., equal current gains in the two stages result in a minimum risetime. In the general case, the optimum F_1 is given by Equation (9.12).*

MULTI-STAGE CIRCUITS

Further improvement of performance may be achieved in some cases by the use of more than two stages. For n stages with identical current

* When the optimization of the individual gains for a *given overall current gain* $F_1 F_2$ is required, Equation (9.12) must be solved numerically or graphically. (See Barna in references.)

gains of $F = I_{DC_n}/I_{DC_{n-1}}$, Equation (9.7) can be extended to

$$t_r = 1.6\tau_0\left\{\left[\left(F + F^n \frac{C\,\Delta V}{\tau_0 I_{DC}}\right)\left(1 + \frac{2r_{b_1}}{R_g}\right)\right]^2 + \sum_{i=1}^{n-1}\left[F + F^{n-i}\frac{C\,\Delta V}{\tau_0 I_{DC_n}}\right]^2\right\}^{1/2}. \quad (9.13)$$

Substituting $C''^* \equiv C\Delta V/\tau_0 I_{DC_n}$, Equation (9.13) becomes

$$t_r = 1.6\tau_0\left\{4[F + F^n C''^*]^2\left[\frac{r_{b_1}}{R_g} + \left(\frac{r_{b_1}}{R_g}\right)^2\right] + \sum_{i=0}^{n-1}[F + F^n F^{-i} C''^*]^2\right\}^{1/2}. \quad (9.14)$$

For a given overall current gain there always exists an optimum number of stages and a corresponding minimum risetime, since it can be shown that

$$t_r \xrightarrow[n\to\infty]{} 1.6\tau_0 F^n C''^*\left\{4\left[\frac{r_{b_1}}{R_g} + \left(\frac{r_{b_1}}{R_g}\right)^2\right] + \sum_{i=0}^{n-1} F^{-2i}\right\}^{1/2} \xrightarrow[n\to\infty]{} \infty. \quad (9.15)$$

The optimum number of stages for a *given overall current gain* can be found analytically when both r_{b_1}/R_g and C''^* can be neglected. In this case it can be shown that the optimum value of n results in a current gain per stage of $F = \sqrt{e} \approx 1.65$.* In the case when only r_{b_1}/R_g can be neglected, the optimum current gain per stage [obtained by evaluating Equation (9.14)] is shown in Fig. 9.2.

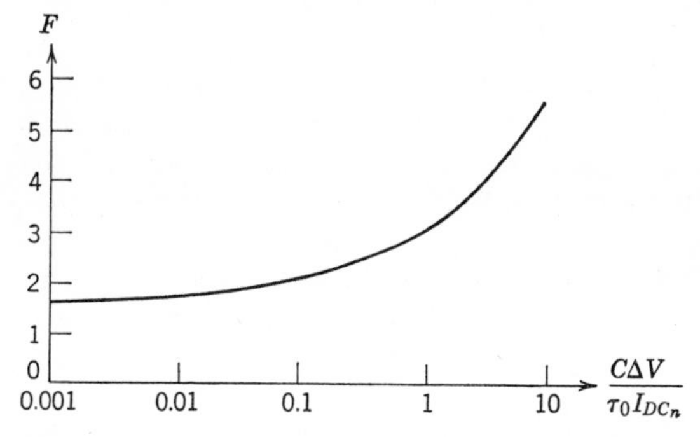

Figure 9.2 Optimum current gain per stage F as function of normalized capacitance $\frac{C\,\Delta V}{\tau_0 I_{DC_n}}$ for multistage digital fanout circuits with $r_{b_1}/R_g \ll 1$.

* This is similar to the results obtained in the optimization of multi-stage linear amplifiers. (See Pettit and McWhorter in references.)

EXERCISES

1. Utilize Equations (9.3) and (9.4) to compute T_0, C^*, R^*, and t_r for a single-stage fanout circuit with $V_G = 250$ mV, $R_g = 50\ \Omega$, $I_{DC} = 20$ mA, $V_T = 30$ mV, $f_T = 1$ GHz, $C = 5$ pF, and $r_b = 10\ \Omega$.

2. Devise a graphical method to compute the optimum F_1 in Equation (9.12) and the risetime given by Equation (9.7) for a two-stage fanout circuit with an input voltage of $V_G = 250$ mV, $R_g = 50\ \Omega$, $I_{DC_2} = 20$ mA, $V_T = 30$ mV, $f_T = 1$ GHz, $C = 5$ pF, and $r_b = 10\ \Omega$.

3. Compute the risetime of a two-stage fanout circuit with the parameters of Exercise 2, but with $F_1 = F_2$. How does this compare to the risetime obtained in Exercise 2?

4. Derive Equation (9.6).

5. Derive Equation (9.9) from Equations (9.7) and (9.8).

6. Prove Equation (9.15) by introducing the overall current gain F^n as a new variable.

7. Prove that for an n-stage fanout circuit with an overall current gain of F^n, the optimum current gain per stage is $F = \sqrt{e}$, if in Equation (9.14) $r_{b1} = 0$ and $C''^* = 0$.

8. Compute the risetime of a three-stage digital fanout circuit with $V_G = 250$ mV, $R_g = 50\ \Omega$, $I_{DC_n} = 20$ mA, $V_T = 30$ mV, $f_T = 1$ GHz, $C = 5$ pF, and $r_b = 10\ \Omega$. How does this risetime compare to the risetime obtained in Exercise 3? What is the optimum number of stages and the corresponding current gain per stage?

9. Compute the risetimes of a two-stage, a three-stage, and a four-stage fanout circuit with the parameters of Exercise 8, but with $I_{DC_n} = 40$ mA.

10. The choice of R_1 and R_2 in the first stage of a two-stage fanout circuit is restricted only by the requirement that the resulting signal be symmetrical with respect to the ground level. In what way will the transient be different if R_1 and R_2 are chosen such that this requirement is observed and if the resulting signal is (a) $\pm V_G/2$ and (b) $\pm 5V_G$?

OOOOO

10

OOOOO

Linear Amplifiers

OOOOOOOOOOOOOOOOOOOOOOOOOOOOOO

The purpose of a linear amplifier is to provide a voltage or current amplification while preserving the shape of the input signal. In this chapter, the transient response and the input impedance of a current amplifier "gain-cell" will be given. This "gain-cell", described by Gilbert (see references), is particularly suited for use in integrated circuit amplifiers.*

THE GAIN-CELL OF GILBERT

A simplified schematic diagram of the amplifier is shown in Fig. 10.1. For zero input current $i_{\text{in}}(t)$, the dc currents in the transistors are determined by I_1, I_{23}, and I_4. Furthermore, for a symmetrical circuit ($I_1 = I_4$, Q_1 identical to Q_4, and Q_2 identical to Q_3), I_{23} splits evenly between Q_2 and Q_3. Small (incremental) signals will be assumed, although Gilbert has shown that the circuit can be linear over a surprisingly large input signal range. Using the hybrid-pi transistor equivalent circuit of Fig. 7.4 with $C_{cb} = 0$ and $\beta \to \infty$,

* For more information on integrated circuit amplifiers see Eimbinder in references.

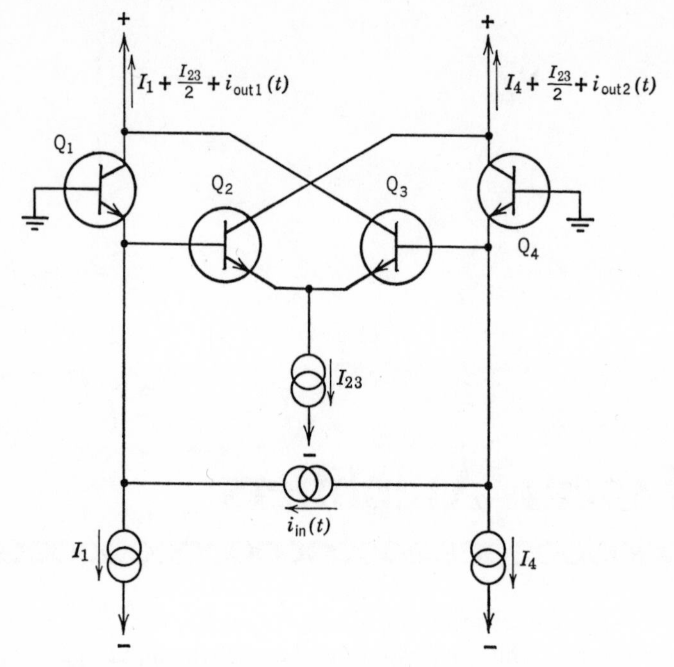

Figure 10.1 Simplified schematic diagram of the gain-cell of Gilbert.

and the input impedance of Fig. 7.6, the amplifier circuit can be drawn as in Fig. 10.2. It can be also shown that for a symmetrical circuit, the signal voltage $v_e(t)$ at the emitters of Q_2 and Q_3 is zero, leading to a simplified analysis.

TRANSIENT RESPONSE

With zero initial conditions, the Laplace transform of i_{D1} can be written (see Exercise 2, Chapter 1, and Exercise 13, Chapter 2) as

$$\mathscr{L}\{i_{D1}\} = \frac{r_{b2} + \dfrac{r_{e2}}{s\tau_0}}{\left(r_{b2} + \dfrac{r_{e2}}{s\tau_0}\right)(1 + s\tau_0) + r_{e1} + r_{b1}s\tau_0} \mathscr{L}\{i_{in}\}, \quad (10.1)$$

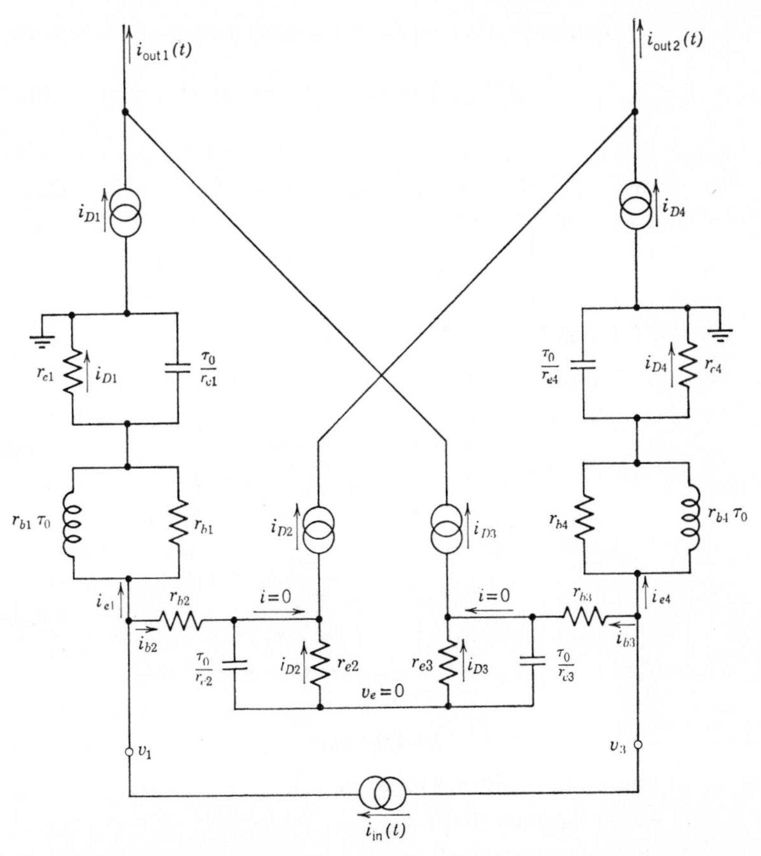

Figure 10.2 Small-signal equivalent circuit of Fig. 10.1 with $\beta \to \infty$ and $C_{cb} = 0$.

and that of i_{b2} as

$$\mathscr{L}\{i_{b2}\} = \frac{r_{e1} + r_{b1}s\tau_0}{\left(r_{b2} + \dfrac{r_{e2}}{s\tau_0}\right)(1 + s\tau_0) + r_{e1} + r_{b1}s\tau_0}\mathscr{L}\{i_{\text{in}}\}. \quad (10.2)$$

From these, the Laplace transform of $i_{\text{out}1}(t)$ can be expressed as

$$\mathscr{L}\{i_{\text{out}\,1}\} = \frac{r_{e2} + r_{b2}s\tau_0 + r_{e1} + r_{b1}s\tau_0}{(r_{e2} + r_{b2}s\tau_0)(1 + s\tau_0) + (r_{e1} + r_{b1}s\tau_0)s\tau_0}\mathscr{L}\{i_{\text{in}}\}. \quad (10.3)$$

In the special case when $r_{e1}/r_{e2} = r_{b1}/r_{b2}$, Equation (10.3) becomes

$$\mathscr{L}\{i_{\text{out}\,1}\} = \frac{G}{1 + Gs\tau_0}\mathscr{L}\{i_{\text{in}}\}, \tag{10.4}$$

where $G \equiv 1 + r_{e1}/r_{e2} = 1 + r_{b1}/r_{b2}$. Thus, the inverse Laplace transform, $i_{\text{out}1}(t)$, for a unit step-function current input, $I_{\text{in}}(t) = u(t)$, can be written as

$$i_{\text{out}\,1}(t) = G(1 - e^{-t/G\tau_0}). \tag{10.5}$$

INPUT IMPEDANCE

The input impedance of the circuit can be written as

$$\frac{\mathscr{L}\{v_1\}}{\mathscr{L}\{i_{\text{in}}\}} = \frac{r_{e1} + r_{b1}s\tau_0}{1 + s\tau_0 + s\tau_0\dfrac{r_{e1} + r_{b1}s\tau_0}{r_{e2} + r_{b2}s\tau_0}}. \tag{10.6}$$

In the special case when $r_{e1}/r_{e2} = r_{b1}/r_{b2}$, Equation (10.6) becomes

$$\frac{\mathscr{L}\{v_1\}}{\mathscr{L}\{i_{\text{in}}\}} = \frac{r_{e1} + r_{b1}s\tau_0}{1 + Gs\tau_0}, \tag{10.7}$$

where $G \equiv 1 + r_{e1}/r_{e2} = 1 + r_{b1}/r_{b2}$ is the same as above.

EXERCISES

1. Derive Equations (10.1) and (10.2).

2. Derive Equation (10.3).

3. Derive Equation (10.4).

4. Sketch the signal of Equation (10.5) with $G = 2$ and $G = 4$.

5. Derive Equation (10.6).

6. Derive Equation (10.7).

7. Utilize the ideas of Figs. 7.6 and 7.7 and design a series compensating network that makes the input impedance of Equation (10.7) resistive. What is the resulting input resistance?

8. Evaluate the component values of the series compensating network of Exercise 7 with $f_T = 1$ GHz, $r_e = 5\ \Omega$, $r_b = 10\ \Omega$, and $G = 2$.

9. What is the 10% to 90% risetime of $i_{\text{out}\,1}$ in the circuit of Exercise 8 for a step-function current input?

10. Modify the risetime obtained in Exercise 9 to include the effect of a 5-pF capacitance in parallel with the input current source $i_{\text{in}}(t)$.

Answers to Selected Exercises

Chapter 1

4. $V = -0.17$ V, $R_g = 2.83\ \Omega$; $V = 0$, $R_g = 3\ \Omega$; $V = 0.17$ V, $R_g = 3.17\ \Omega$

Chapter 2

5. $V_L(t) = V_0 e^{-Rt/L}$; $V_R(t) = V_0(1 - e^{-Rt/L})$

6. $I_C(t) = I_0 e^{-t/RC}$; $I_R(t) = I_0(1 - e^{-t/RC})$

7. $I_L(t) = I_0(1 - e^{-Rt/L})$; $I_R(t) = I_0 e^{-Rt/L}$

Chapter 3

1. $\dfrac{LCs^2}{1 + RCs + LCs^2}$

Chapter 4

1. 0.12 Ω

6. 50 Ω

7. 350 Ω

8. 12 inches; 4000 feet

Chapter 5

1. ≈ -10 nA; ≈ -10 nA; 0; 3 mA; 90 mA

3. 65 mV

4. $10^{264}\ \Omega$, 10^{-274} F; $3.3 \times 10^{33}\ \Omega$, 3×10^{-43} F; $3.7 \times 10^{6}\ \Omega$, 2.6×10^{-17} F; 12.5 Ω, 8×10^{-12} F; 0.4 Ω, 2.5×10^{-10} F

6. 0.26 pF; 0.62 pF; 1 pF; 1.7 pF; 2.2 pF

7. 40 ps; 70 ps

Chapter 6

2. $V \cong 2\,V_p$; $r_i I_p / V_p \cong -e$
3. -54 ps; 22 ps
4. 0.4 ns
7. 0.45 ns, 0.5 ns, 0.55 ns; $C_2 = 1$ pF

Chapter 7

10. 3 pF, 100 Ω; 30 pF, 500 Ω; 100 Ω to 1 kΩ

Chapter 8

6. 1 ns; 1.1 ns
7. 1.05 ns, 1.15 ns; 1.2 ns, 1.3 ns
8. 0.3 ns, 1.5 ns

Chapter 9

1. $T_0 = 1$ ns, $C^* = 1.35$, $R^* = 1.4$, $t_r = 1.9$ ns

Chapter 10

8. 5 Ω, 50 pF, 5 Ω, 1.6 nH
9. $t_r = 0.25$ ns
10. $t_r = 0.27$ ns

References

General References

R. Littauer, *Pulse Electronics*, McGraw-Hill, New York, 1965.

J. Millman and H. Taub, *Pulse, Digital, and Switching Waveforms*, McGraw-Hill, New York, 1965.

Chapter 1

R. S. Smith, *Circuits, Devices, and Systems*, John Wiley and Sons, New York, 1966.

Chapter 2

T. M. Apostol, *Mathematical Analysis*, Addison-Wesley, Reading, Mass., 1957.

R. Bracewell, *The Fourier Transform and Its Application*, McGraw-Hill, New York, 1965.

A. Erdelyi et al., *Tables of Integral Transforms*, Vol. 1., McGraw-Hill, New York, 1954.

F. E. Nixon, *Handbook of Laplace Transformation*, Prentice-Hall, Englewood Cliffs, N.J., 1965.

G. E. Roberts and H. Kaufman, *Table of Laplace Transforms*, W. B. Saunders Company, Philadelphia, 1966.

Chapter 3

W. C. Elmore, "The Transient Response of Damped Linear Networks with Particular Regard to Wideband Amplifiers," *Journal of Applied Physics*, **19**, 55–63 (January 1948).

W. C. Elmore and M. Sands, *Electronics, Experimental Techniques*, McGraw-Hill, New York, 1949.

F. F. Kuo, *Network Analysis and Synthesis*, John Wiley and Sons, New York, 1962.

M. E. Van Valkenburg, *Network Analysis*, Prentice-Hall, Englewood Cliffs, N.J., 1964.

Chapter 4

R. E. Matik, *Transmission Lines for Digital and Communication Networks*, McGraw-Hill, New York, 1969.

P. M. Morse and H. Feshbach, *Methods of Theoretical Physics*, McGraw-Hill, New York, 1953.

R. L. Wigington and N. S. Nahman, "Transient Analysis of Coaxial Cables Considering Skin Effects," *Proc. IRE*, **45**, 166–174 (February 1957).

Chapter 5

A Barna and D. Horelick, "A Simple Diode Model Including Conductivity Modulation," *IEEE Transactions on Circuit Theory* (In press).

J. F. Gibbons, *Semiconductor Electronics*, McGraw-Hill, New York, 1966.

W. H. Ko, "The Forward Transient Behavior of Semiconductor Junction Diodes," *Solid State Electronics*, **3**, 56–69 (1961).

J. G. Linvill, *Models of Transistors and Diodes*, McGraw-Hill, New York, 1963.

J. Milman and C. C. Halkias, *Electronic Devices and Circuits*, McGraw-Hill, New York, 1967.

R. L. Pritchard, *Electrical Characteristics of Transistors*, McGraw-Hill, New York, 1967.

Chapter 6

A. Barna, "An Analysis of Transients in Tunnel-Diode Circuits," Report SLAC-88, Stanford Linear Accelerator Center, Stanford University (August 1968).

W. F. Chow, *Principles of Tunnel Diode Circuits*, John Wiley and Sons, New York, 1964.

A. Ferendeci and W. H. Ko, "A Two-Term Analytical Approximation of Tunnel Diode Static Characteristics," *Proc. IRE*, **50,** 1852–1853, (August 1962).

J. O. Scanlan, *Analysis and Synthesis of Tunnel Diode Circuits*, John Wiley and Sons, New York, 1966.

Chapter 7

C. L. Alley and K. A. Atwood, *Electronic Engineering*, John Wiley and Sons, New York, 1966.

H. N. Ghosh, F. H. DeLaMoneda, and N. R. Dono, "Computer-Aided Transistor Design, Characterization, and Optimization," *Proc. IEEE*, **55,** 1897–1912, (November 1967).

D. L. Schilling and C. Belove, *Electronic Circuits: Discrete and Integrated*, McGraw-Hill, New York, 1968.

Chapter 8

A. Barna, "High-Speed Switching Properties of the Emitter-Coupled Transistor-Pair," Report SLAC 97, Stanford Linear Accelerator Center, Stanford University (March 1969).

R. D. Middlebrook, *Differential Amplifiers*, John Wiley and Sons, New York, 1963.

Chapter 9

A. Barna, "Risetime Optimization of Digital Fanout Circuits," *IEEE Journal of Solid-State Circuits*, **SC-4,** 159–161 (June 1969).

J. M. Pettit and M. M. McWhorter, *Electronic Amplifier Circuits*, McGraw-Hill, New York, 1961.

Chapter 10

J. Eimbinder, *Designing with Linear Integrated Circuits*, John Wiley and Sons, New York, 1969.

B. Gilbert, "A New Wide-Band Amplifier Technique," *IEEE Journal of Solid-State Circuits*, **SC-3,** 353–365 (December 1968).

Index